RATIOS

PROPORTIONS

PROGRESSIONS

&

SPIRALS

By Arthur Yuenger

MATHEMATICS IS AN ANTIDOTE FOR DOGMA

(anonymous)

Table of Contents

Introduction

During the summer of 1980 or thereabouts, I was living in Santa Barbara, California. On one of the typically sunny days, I happened to be walking down State Street, the main thoroughfare lined with shops. One such curio shop had a chambered nautilus shell displayed in the window. It had been sawn in half to reveal its inner structure. I was immediately struck by its exquisite geometry: I went inside and purchased the shell.

When I returned to my apartment, I quickly went to my desk, removed the shell from the clerk's protective packaging, and sat there somewhat mesmerized by its beauty. It was apparent that the shell is an expression of mathematics. At that moment I knew I had to find out exactly what that mathematics is.

During the next several years I came to understand that the cross section of the nautilus shell is a near perfect representation of the definition of a logarithmic spiral, and that the algebraic polar equation for the curve is:

$$r = e^{0.18314\theta}$$

I also discovered that its graceful curve is determined by a simple ratio of four to three (4/3). This relationship enables the equation above to be written as a function of this ratio:

$$r = (4/3)^{2\theta/\pi}$$

Sometime during the 1990's, after leaving California to return to Colorado, the idea came to me that perhaps the profile of chicken eggs might also be structured according to a simple ratio. I purchased a dozen eggs at the supermarket, and proceeded to measure the lengths (the longest axial dimension) corresponding to the widths (the widest, transverse dimension). Lo and behold, the average ratio of length corresponding to diameter was 4/3. This result prompted me to look into the proportions of other bird eggs

As it happened, there was a ranch near Rifle, Colorado where emus were raised (birds resembling small ostriches). I stopped by one day, introduced myself to the rancher, explained that I was an architect interested in natural proportions, and asked if I could measure some of his emu eggs. He said "sure, no problem." He had several empty shells in the ranch house. The shells were whole; the inside contents had been liquefied and removed through a small hole that had been drilled at one of the ends. I measured several eggs: the average length to width ratio was 3/2.

This got me to thinking about the proportions of ostrich eggs. I managed to locate a ranch near Cortez, Colorado where ostriches were being raised. I phoned the owner to explain my interest. He too was very friendly, and invited me to visit. I measured several eggs, and found that the average ratio of length to width was 5/4.

Remarkably, the proportions of these three different species of eggs are also the terms of a sequence of fractions beginning with 2/1:

$$\frac{2}{1}; \ \frac{3}{2}; \ \frac{4}{3}; \ \frac{5}{4}; \ \frac{6}{5}; \ \frac{7}{6}; \ \frac{8}{7}; \ \frac{9}{8}; \ \frac{10}{9}; \ \frac{11}{10} \ ...$$

At the time I discovered this relationship, I became acquainted with a branch of biology referred to as phylotaxis: the study of the arrangement of leaves on the stems of plants. It turns out that the leaves of numerous plants are positioned on stems according to simple ratios. The ratios relate to a specific number of leaves corresponding to a specific number of invisible spiral lines that wrap around the stem. What is remarkable is that the leaf/spiral ratios are related to a famous sequence of numbers referred to as the Fibonacci sequence:

$$0, 1, 1, 2, 3, 5, 8, 13, 21, 34, 55, 89, \ ...$$

Organized sequences of numbers are an important aspect of mathematics. A fascinating example is a multiplication table structured by a horizontal and vertical array of arithmetic progressions:

1	2	3	4	5	6	7	8	9	10
2	4	6	8	10	12	14	16	18	20
3	6	9	12	15	18	21	24	27	30
4	8	12	16	20	24	28	32	36	40
5	10	15	20	25	30	35	40	45	50
6	12	18	24	30	36	42	48	54	60
7	14	21	28	35	42	49	56	63	70
8	16	24	32	40	48	56	64	72	80
9	18	27	36	45	54	63	72	81	90
10	20	30	40	50	60	70	80	90	100

The table shown above is composed of ten rows of arithmetic progressions beginning with the progression 1, 2, 3, 4, 5, 6, 7, 8, 9, 10, and ending with 10, 20, 30, 40, 50, 60, 70, 80, 90, 100. The product of any number in the top row multiplied by any number in the far left column is equal to the number that lies at the intersection of the column and the row containing the multipliers. For example, the product of 7 in the left column and 8 in the top row is 56: the number at the intersection of the row beginning with 7, and the column beginning with 8.

This example of a multiplication table resulting from a simple array of arithmetic progressions; together with my discovery postulating the proportions of bird eggs and seashells as related to a sequence of simple fractions or Fibonacci numbers; together with my understanding of the ratios of phylotaxis as related to the Fibonacci sequence; together with my analysis of the mathematics of the logarithmic spiral and its application to the geometry of the shell of the chambered nautilus—indicates that there is an intelligence at work in the Universe. What is most impressive, if not astounding, is the magical manner in which this intelligence, in the form of simple ratios and sequences, expresses itself.

The first half of this book will illustrate several expressions of this intelligence beginning with a comprehensive review of the principles of ratios, proportions, sequences and progressions. The presentation will include a discussion of the most famous ratio of all—the golden ratio—and a comprehensive review of the mathematics of the logarithmic spiral. The last section includes an analysis of the geometry of the chambered Nautilus as related to the logarithmic spiral. The second half of the book, the appendix, illustrates derivations of mathematic principles relating to discussions in the first half.

It was never intended that this book would be a mathematical text. The mathematics presented occurred as a result of my inquiry into the geometry of familiar natural forms. Although much of my high school and college math had become dormant, but not forgotten—except for calculus—I was able to recall what I had learned in order to understand the various derivations and relationships associated with natural or geometric forms. All of the steps required to derive formulas shown in this book (including those steps that are obvious to mathematicians) are incorporated in the various derivations of formulas and relationships.

Over the course of the time that it took to develop and prepare this text, I was very fortunate to come in contact with various individuals whose knowledge of mathematics exceeded mine, and who provided me with the mathematics I needed to know at the time I needed to know it. For example, when I began my inquiry into the structure of the nautilus shell while living in Colorado, I drove to Fairfield, Iowa to visit some friends for a week. I was able to offer a ride from Denver to Iowa City, to Bruce Simon, Ph.D. candidate in biological physics at the University of Iowa. During our many conversations, I mentioned that I had discovered a relationship between logarithmic spirals and simple ratios. I explained that I needed an equation for the spiral as related to a ratio. In a matter of a few minutes, Bruce went through the steps to produce the equation (which, after having refreshed my memory regarding the nature of logarithms, I was able to simply). At another time, while I was living in Fairfield, another individual, Jeff Blocker, who had a degree in electrical engineering, derived the equations for determining the area and length of a portion of a logarithmic spiral.

Above all was Charles Campbell, physicist and ophthalmic consultant. Charlie and I were fraternity brothers at the University of Illinois during the years 1957 through 1961. He is a brilliant mathematician and inventor who enjoys intellectual challenges. I contacted him a few years ago, and he was more than delighted to help with the derivations of the formulas for the area and length of the Pentagram Koch curve.

Anyone with a rudimentary grasp of mathematics, that is to say, anyone who understands the purchase of goods at the supermarket, should find this book informative even if some of the mathematics presented is not understood. Conversely, anyone who is a mathematical wizard with a complete understanding of all aspects of mathematics should find this book a pleasant review.

Arthur Yuenger
Fairfield, Iowa
2019

Ratios and Proportions

One of the most fundamental principles of mathematics is that of a ratio: the comparison of two things in terms of quantity, degree or size. Almost everyone engages in this simple principle on a daily basis. For example, the measurement of the length of an object is a ratio comparing a specific quantity of unit lengths to a standard unit length. Similarly, the weight of an object is also expressed according to a ratio: the comparison of its weight to a standard unit of weight.

A baseball player's batting average is a ratio representing the number of hits made by the player corresponding to the number of times the player appeared at bat, indicated by a decimal number expressed in thousandths. For example, if the player has scored 21 hits in 63 times at bat, his average is 21/63 = 0.333, and he is said to be batting *three-thirty-three.*

The player's team standing is also represented as a ratio, also expressed in thousandths. If his team has won 19 out of 32 games played, the team's average is 19/32 = 0.594 (*five-ninety-four*). If the team with the most wins has won 23 out of 32 games, its average is 0.720, and is considered to be in first place. The player's team is considered 4 games out of first place, meaning that his team must add 4 winning games to the fraction determining its ratio of games won to games played, in order to equal the ratio (standing) of the first place team. (See Appendix A.)

Anyone who drives a vehicle knows that speed is determined according to a ratio expressed as a rate: the comparison of one quantity corresponding to another unit quantity of something else—distance traveled as related to a unit of time such as miles (or kilometers) per hour. Fuel for the vehicle is also purchased according to a ratio expressed as a rate: cost as related to unit quantity, which is to say, currency as related to a unit volume of fuel such as dollars per gallon. The amount of air in the tires is also determined according to a rate: the pressure or force of the air in the tire corresponding to the unit area upon which it acts such as pounds per square inch (p.s.i.).

When space is leased in a building, rent is paid according to the floor area of the rented space multiplied by a ratio representing the total area of a floor corresponding to the sum of the area of rentable space on the floor. Many municipalities limit the size of a building according to a Floor Area Ratio: the sum of the area of all the floors (measured in a prescribed manner) as related to the area of the building lot.

Laboratory experiments have shown that the deflection Δ, at the center of a uniformly shaped structural beam, induced by a load P at its center, is equal to a ratio of the imposed load multiplied by the length L of the beam raised to the power of three, corresponding to forty-eight times the modulus of elasticity E, multiplied by the moment of inertia I, of the beam:

$$\Delta = \frac{PL^3}{48EI}$$

When we listen to music, we listen to distinct combinations of tones arranged according to specific ratios comparing the frequencies of the tones. The frequencies correspond to vibrations of molecules, and are expressed as rates referred to as cycles per second.

The most fundamental relationship involving shapes and forms is that of a proportion: the comparative relation between any two aspects of the various dimensional attributes of an object. In the context of mathematics, a proportion is defined as the equality of two ratios. For example, any four quantities represented by the letters a, b, c, and d, arranged such that a divided by b is equal to c divided by d defines a proportion:

$$\frac{a}{b} = \frac{c}{d}$$

This relationship is also written $a{:}b = c{:}d$ (read a is to b as c is to d). The terms b and c are referred to as the means, and the terms a and d are referred to as the extremes. It is easily shown that the product of the means is equal to the product of the extremes (see appendix B):

$$b\times c = a\times d$$

The use of proportional ratios occurs regularly whenever anyone is involved with shopping for groceries. For example, with regard to a portion of a receipt from a grocery store,

```
PRODUCE [NON TAX]
0.71 lb @ $1.19/ lb
POTATO - RUSSET ORGANIC          $0.84 F
0.45 lb @ $2.99/ lb
PEPPER - GREEN ORGANIC           $1.35 F
CHARD - RAINBOW ORGANIC          $2.19 F
0.65 lb @ $2.99/ lb
PEPPER - RED ORGANIC             $1.94 F
1.65 lb @ $1.05/ lb
BANANAS - ORGANIC                $1.73 F
4 @ $0.35 each
KIWI - ORGANIC                   $1.40 F
2 @ $1.09 each
AVOCADO - ORGANIC                $2.18 F
```

seventy-one hundredths of a pound of russet potatoes were purchased at a rate of one pound corresponding to one dollar and nineteen cents, resulting in a total cost of eighty-four cents. The eighty-four cents is actually derived from a proportional ratio; x is to *0.71* pounds as *$1.19* is to *1* pound, where x is the cost of the purchased potatoes, such that $x = 1.19 \times 0.71 = \$0.84$:

$$\frac{x}{0.71 \; lb.} = \frac{\$1.19}{1.00 \; lb.}$$

The cost of the other articles listed is determined in the same manner. Indeed, most bulk merchandise sold in any marketplace is purchased in a similar manner.

It is readily apparent that the fundamental mathematical concept of ratios and proportions affects all aspects of life. This organizing quality of intelligence enables anyone to understand and communicate relations between things in a systematic and orderly manner.

What is most fascinating is the way in which the basic principle of simple ratios and proportions is expressed in natural forms, and the degree to which it is associated with aesthetics. Consider for example, a bird's egg. This simple form has been described as nature's perfect package, expressing complete harmony between form and function. Not only does this shape offer maximum protection of its contents from external forces such as the weight of the parent bird, but in a completely opposite manner, the very same shape offers minimum resistance from internally applied forces, enabling the chick to easily peck its way to freedom.

However, it is the shape and not the function of the shape that is so very appealing. The egg is also one of nature's most elegant forms, invoking pleasing emotions as it demonstrates a clear example of symmetry, harmony and balance. It also expresses a clear, albeit elementary, example of the mathematical definition of a proportion: the diameter of the maximum circular girth is related to the length according to a ratio.

As mentioned in the introduction, the measurements of a few eggs of three different species of birds suggest that the average ratio of the diameter of an egg corresponding to its length is expressed by specific whole numbers. (See the following photographs of an emu, chicken and ostrich egg.) Remarkably the length appears to be one more unit than the width. For example, if the diameter of the girth of an egg is 2 units, then the length is 3 units; if the diameter is 3, the length is 4 and so forth. These ratios can be arranged according to a simple sequence of fractions beginning with 1/2:

$$\frac{1}{2}; \ \frac{2}{3}; \ \frac{3}{4}; \ \frac{4}{5}; \ \frac{5}{6}; \ \dots$$

The emu egg illustrated above is shown at approximately the same scale as the chicken egg illustrated below. The proportions of the emu egg, (the diameter of the girth corresponding to the length of the egg) are two to three. The actual dimensions are 3.73 × 5.55 inches. The proportions of the chicken egg (girth to length) are 3 to 4. The actual dimensions of the chicken egg are 1.72 × 2.35 inches.) Note that the girth of all eggs is a perfect circle.

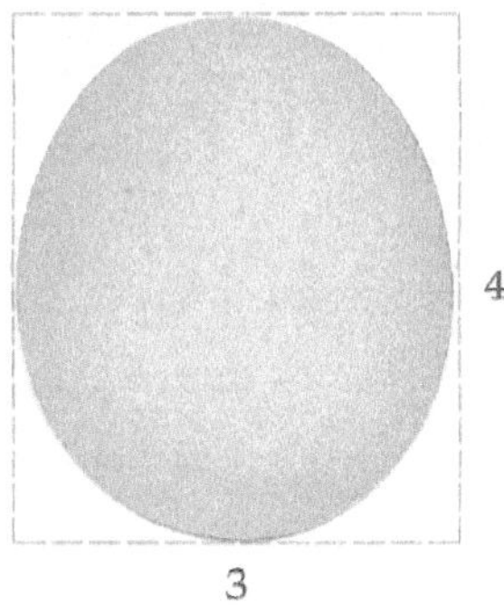

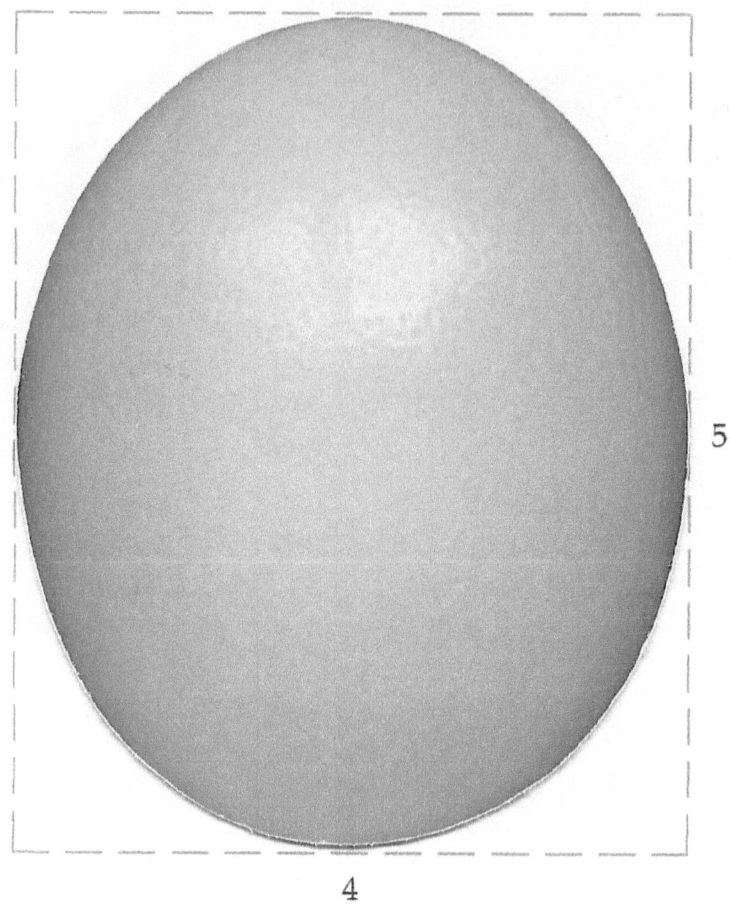

The ostrich egg illustrated above is shown at approximately the same scale as the chicken and emu egg that appear on the previous page. The proportions of the ostrich egg (girth diameter to egg length are 4 to 5. The actual dimensions are 4.96 × 6.20 inches.)

See Appendix Q for my proposed geometric construction of the profiles of ostrich, emu and chicken eggs.

Ratios and the Fibonacci Sequence

A sequence is defined as an ordered succession of terms arranged according to a rule. For example, the sequence of ratios,

$$\frac{1}{2};\ \frac{2}{3};\ \frac{3}{4};\ \frac{4}{5};\ \frac{5}{6};\ \frac{6}{7};\ \frac{7}{8};\ \frac{8}{9};\ \frac{9}{10};\ \frac{10}{11};\ \frac{11}{12};\ \dots$$

is established by the rule that beginning with 1/2, the numerator of each successive term is equal to the denominator of the previous term. The denominator of each successive term is equal to the denominator of the preceding term plus one. The sequence can be continued indefinitely, with the decimal value of each term approaching, but never equaling, unity:

0.50; 0.666...; 0.75; 0.80; 0.833...; 0.875; 0.90; 0.909...; 0.916...

Yet another sequence of ratios, (similar to the above succession in that the rule establishing the numerators is the same) is the following:

$$\frac{1}{2};\ \frac{2}{3};\ \frac{3}{5};\ \frac{5}{8};\ \frac{8}{13};\ \frac{13}{21};\ \frac{21}{34};\ \frac{34}{55};\ \frac{55}{89};\ \frac{89}{144};\ \dots$$

In this sequence (also beginning with 1/2) the denominator of each succeeding term is also equal to the sum of the numerator and denominator of the previous term. The numerator is equal to the denominator of the previous term. This sequence can also be extended indefinitely, with the value of each term approaching (to five decimal places), 0.61803... :

0.50; 0.666...; 0.60; 0.625; 0.615...; 0.619...; 0.6176...; 0.61818...;

The most remarkable feature of this sequence is that the integers comprising each term of the numerator and denominator are the integers of the famous Fibonacci sequence:

0, 1, 1, 2, 3, 5, 8, 13, 21, 34, 55, 89, 144, 233, ...

The Fibonacci sequence is formed with the rule that beginning with 0 and 1, each successive term is the sum of the previous two terms:

$$0 + 1 = 1, \quad 1 + 1 = 2, \quad 2 + 1 = 3, \quad 3 + 2 = 5, \text{ and so forth.}$$

Although the sequence begins with zero, Fibonacci numbers are contained in any sequence formed by this rule irrespective of the starting numbers. For example, beginning with any numbers represented by the letters a and b, the sequence becomes:

$$a, \quad b, \quad a + b, \quad a + 2b, \quad 2a + 3b, \quad 3a + 5b, \quad 5a + 8b, \text{ and so forth.}$$

The sequence is named after Leonardo Bonacci, the brilliant mathematician of the twelfth century. Fibonacci—a contraction of fillious Bonacci (son of Bonacci)—was born in Pisa, Italy about 1175 AD. He is credited with the publication of *Liber Abaci* which in 1202 AD, introduced the Arabic (decimal) system of numeration to the intelligentsia of Christian Europe who were, at that time, using the cumbersome Roman numeral system. It is thought that Fibonacci learned the decimal system while living with his merchant father at Bugia, in Barberry, North Africa (now Bejaia, Algeria).

Fibonacci numbers are special for a variety of reasons, the most important of which are (1) they relate to the golden proportion, and (2) they occur frequently, almost ubiquitously, in the expression of nature's intelligence. These numbers account for not only sequences involving spiral patterns in plants and animals, but also the proportions of the various forms of nature, and the arrangement of their constituent organs.

For example, the overlapping scales of pine cones are arranged according to spiral rows originating at the base of the cone (where it is attached to the stem), and ending at the tip. The spiral rows are called *parastichies.* They occur as opposing sets established by Fibonacci ratios. The spiral rows of petals on the cone shown on the next page are arranged according to the 5/8 ratio (Fibonacci numbers 5 and 8). Five spirals of petals wrap around the central stem in a counterclockwise direction, and eight spirals wrap around in a clockwise direction. With regard to the pine cone shown, one of the five spirals has been painted white:

Correspondingly, eight spirals wrap around the stem of the pine cone in the opposite direction. One of the eight spirals, indicated by the arrows, has been painted black. Other ratios that determine the parastichial arrangement of the scales of other species of cones are 2:3, 3:5, and 8:13.

Additionally, the proportions of the cone (with scales extended as shown) are expressed according to a simple proportional ratio: length *l* is to width *w* as three is to one. (*l:w = 3:1*).

If the fleshy portion of the cauliflower plant is carefully removed by cutting the stems supporting the florets from the conical stalk, an arrangement of parastichies according to a 3:5 ratio is revealed:

The stems supporting the florets are positioned according to five spirals rotating in one direction (one of which is indicated by the darker glass beads placed on the ends of the cut stems), and three spirals rotating in the opposite direction (one of which is indicated by the lighter beads placed on the ends of the cut stems). Similar arrangements of spirals can be found in other plants of the mustard family such as cabbages, broccoli and Brussels sprouts.

* * * * * * * * * * *

The seeds forming the disc portion of flower blossoms are also arranged according to parastichies. Sunflowers, with such large discs are good examples. The disc on the sunflower on the next page contains 89 spirals rotating clockwise and 55 spirals rotating counterclockwise, producing a 55:89 ratio. (At the time of the photograph, the petals forming the corolla had dried and twisted revealing the underlying sepals. The florets attached to the ends of the husks of the seeds were also in the process of drying. Some had fallen off causing the lighter portions on the disc.)

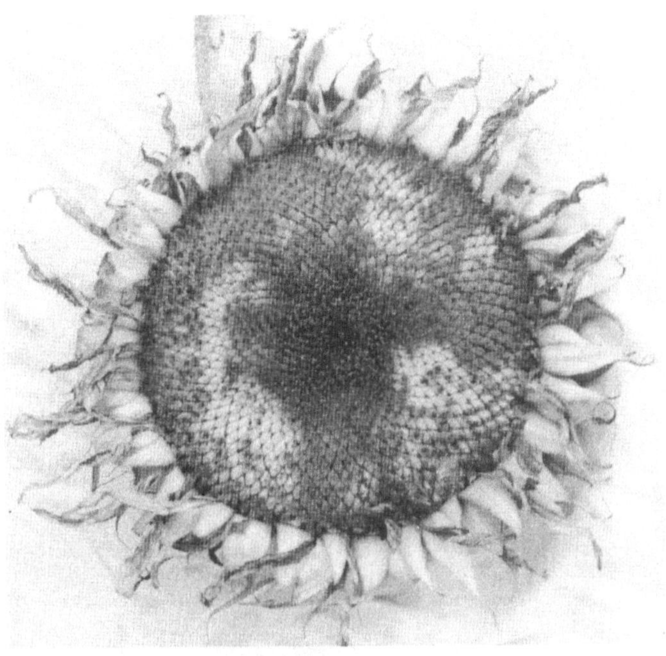

In the illustration below of a completely dried sunflower disc, 68 spirals radiate clockwise, and 42 radiate counterclockwise forming a 42:68 ratio, which when reduced by one-half equals the Fibonacci ratio 21:34. (The seeds of one row of each of the opposing sets of spirals have been removed.)

The parastichies arranging the bracts on the surface of a pine-apple occur in sets according to an 8:13 ratio.

* * * * * * * * * * *

Not only are the spiral arrangements of plant organs related to Fibonacci numbers, but so are the constituent parts. For example, the number of petals forming the blossom of a flower will often (but not always) total a Fibonacci number. Daisies usually have either 13 or 21 petals, whereas the black-eyed Susan averages 13:

Additionally, the center disc containing the seeds of this black-eyed Susan has a diameter of 3 units, while the tips of the 13 petals forming the corolla fit inside a circle whose diameter is 13 units. Further, using the same Fibonacci units of measurement, each maximum petal width is 2 and each petal length, measured from the edge of the disc to the tip of the petal is 5.

* * * * * * * * * * *

Fibonacci numbers also appear in seashells. The photographs on the next page illustrate two views of the shell of a bivalve mollusk known as a cockle. The upper photograph (looking down on one of the shell halves) shows the ribbed convex surface. There are 34 radial ribs on each of the two halves of the shell. This exquisite pattern can be counted in the end view (lower photograph). Although not easily discernible, the first rib on the upper shell, appearing as a slight bulge, begins directly above the darker portion of the junction between the shell halves, just to the left of the center portion of the ribs, below the clear arc identifying the edge of the second rib.

The proportions of the smallest rectangle into which the profile can be inscribed are 4:5:

The profile, from an end view of the two halves, can be inscribed in a square. The upper and lower portions of the curves of the halves can be inscribed in a circle inscribed in the square:

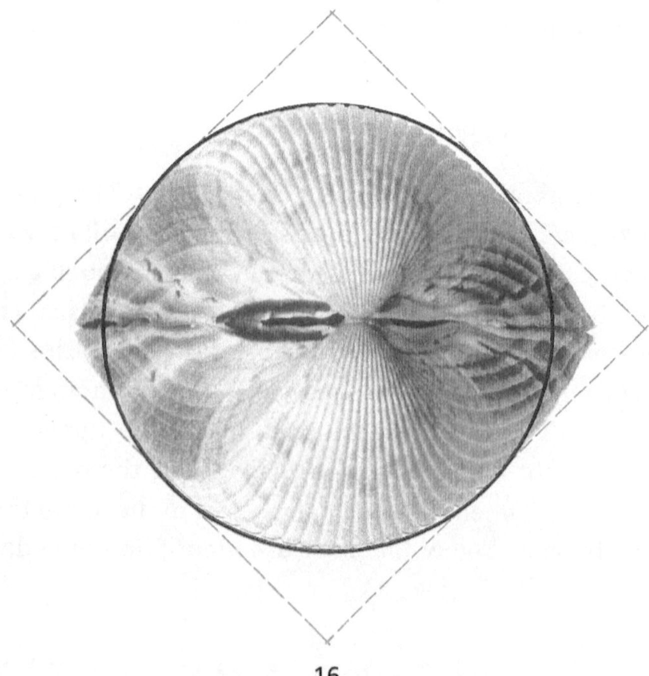

Phyllotaxy

Phyllotaxy (also called phyllotaxis) refers to the arrangement of leaves on the stems of plants. As a plant grows, the emergence of its leaves does not occur in a random, haphazard manner. Each leaf is positioned according to simple yet precise geometric rules. One such set of rules relating to a variety of plants involves coiled spirals and the terms of the Fibonacci sequence expressed as ratios:

$$\frac{1}{2}; \ \frac{1}{3}; \ \frac{2}{5}; \ \frac{3}{8}; \ \frac{5}{13}; \ \frac{8}{21}; \ \frac{13}{34}; \ \ldots$$

If a continuous invisible line is drawn connecting the point of attachment of leaves to the stem of certain plants, the line will form a helical spiral coiling around the stem. There will always be a set quantity of leaves equal to a Fibonacci number corresponding to successive coils of the spiral which are also equal to a Fibonacci number. Both numbers are expressed in one of the ratios of the sequence illustrated above.

The numerators of these ratios represent the number of successive coils of the spiral. (One coil is equivalent to one complete rotation of 360°.) The denominators represent the number of leaves between the starting and ending points of the spirals. The fractions (ratios) establish the direction of the projection of each leaf from the stem according to equal segments of a circle established by equally spaced radial lines. The radial lines are drawn on an invisible, imaginary plane perpendicular to the base of the stem. The angular spacing of the radial lines is also established by the denominators of the fractions.

Phyllotaxy is best understood by examining the leaf arrangements of a few familiar plants that conform to these rules. For example, with regard to a common grass plant, the arrangement of sessile leaves (leaves attached directly to the stem) is structured by the Fibonacci ratio 1/2. The rules establishing the emergence of the leaves are illustrated in the following sketch:

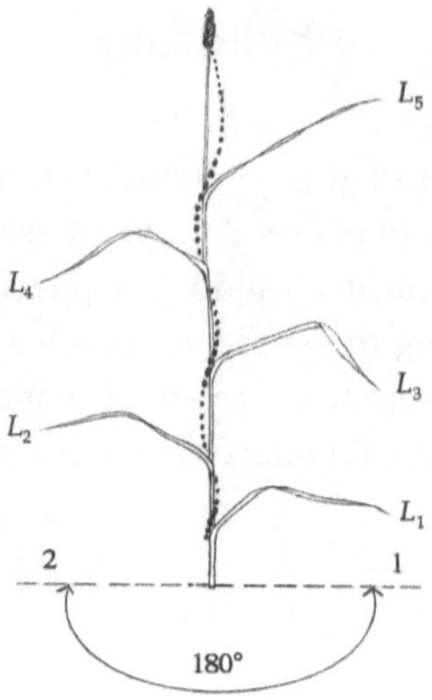

The 180° arc on the imaginary invisible plane at the base of the
stem is drawn as if viewed in perspective.

Invisible, imaginary radial lines 1 and 2 are drawn on the invisible plane perpendicular to the base of the stem. The radial lines
are spaced according to an arc of 180° (360°÷2). The radial lines
represent the directions of the projection of each leaf from the
stem. The invisible line of the spiral, coiling counterclockwise in
this example (as it winds upward around the stem), begins at the
attachment of leaf (L_1). The next leaf (L_2) projects directly above
and in the same direction as radial line 2 completing one revolution
of the spiral such that two leaves (L_1 and L_2) correspond to one
complete coil of the spiral. The next leaf (L_3) projects directly above
and in the same direction as leaf (L_1).

The sketch on the next page illustrates a 2:5 leaf arrangement
of a yarrow plant stem supporting its corymbed (flat-topped) inflo-
rescence. Five equally spaced radial lines (1, 2, 3, 4, and 5) have
been drawn on the invisible plane at the base of the stem, each pro-
jecting outward to indicate the direction of the clusters of sessile,
pinnate (resembling feathers) leaves attached to the stem. Since
this is a 2:5 arrangement, the spacing of the radial lines is 360° ÷ 5

= 72°. The spacing of the radials representing the projection of the leaves is 360° × 2/5 = 144° (or 72° × 2 = 144°).

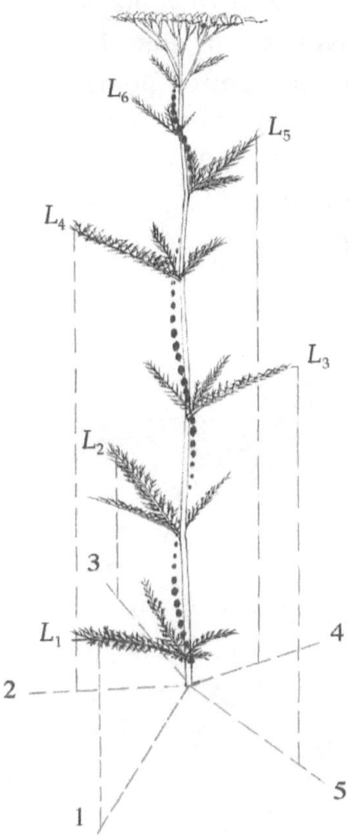

The lowest cluster of leaves (L_1) is assigned to the first radial line. The invisible spiral drawn dotted (coiling clockwise in this example as it winds upward around the stem) begins at the attachment of this group of leaves to the stem. The next leaf cluster (L_2) projects in the same direction as radial line 3, 144° from radial line 1. The third leaf cluster (L_3) projects in a direction corresponding to radial line 5, spaced 144° from radial line 3 completing one coil of the spiral. The fourth cluster (L_4) aligns directly over radial line 2, and the fifth cluster (L_5) directly above radial line 4 completing the second coil of the spiral, such that five ranks of leaf clusters correspond to two coils of the spiral. The sixth leaf cluster (L_6) aligns once again over the first radial line,

Yet another example of phyllotaxy is illustrated by a four-month growth of leaves that produced a new stalk on one of two main stems, A and B, of a potted poinsettia plant. The leaves arranged themselves according to a 3:8 ratio as illustrated by the following sketch. The view is looking down from a point just above the leaves, showing the stems A and B in section. The new stalk has emerged from stem B.

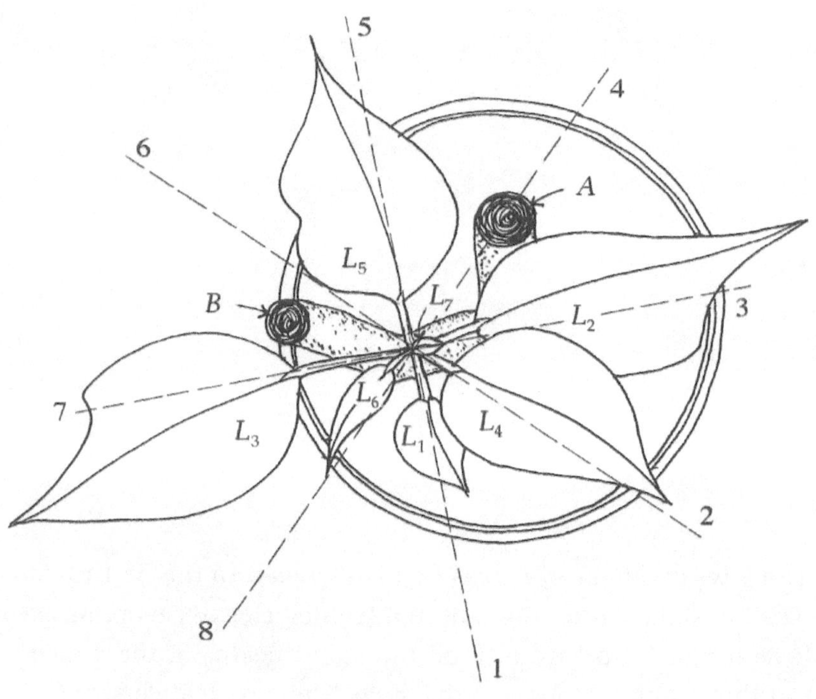

Since eight leaves will grow according to three spirals (coiling counterclockwise) on the new stalk, the radial spacing of the leaves will be 3/8 × 360° = 135°. The equal spacing of the radial lines is 360° ÷ 8 = 45°. The petioles (slender stems supporting the leaves) will align on every third radial line (3 × 45°). As in the previous examples, the first leaf (L_1) to have emerged is assigned to the first radial line 1. (At the time the sketch was drawn, this leaf had not fully matured, and remained smaller than the other leaves.)

The second leaf (L_2) that emerged, should have aligned on radial line 4 (135° from radial line 1), but did not because stem *A* was in its way. The petiole of the third leaf (L_3) did align directly on radial line 7, three equal segments from radial line 4. Likewise, the fourth leaf (L_4) aligned on radial line 2 (completing one and one-eighth coils of the spiral). The fifth leaf (L_5) aligned on radial line 5, and the sixth leaf (L_6) aligned on radial line 8. The tiny seventh leaf (L_7), almost indiscernible, had just emerged from the end of the new stem, preparing to project in a direction along radial line 3, thus completing two and one-quarter coils of the spiral.

The eighth leaf that had not yet emerged will grow in a direction along radial line 6. The ninth leaf which also had not yet emerged will appear directly above the first leaf, aligning once again on the first radial line, completing the rank of 8 leaves corresponding to 3 spirals. Note that the red bracts (leaves resembling petals) for which the plant is famous, occurred at the ends of stems *A* and *B.*

The sequential arrangement of leaves on the stems of plants according to Fibonacci numbers has a purpose. It creates uniform spacing between the leaves, both vertically and horizontally, configuring the plant as a balanced whole to ensure proper functioning of the entire organism.

Progressions

An **arithmetic progression** is a sequence of numbers whose common difference between consecutive terms is constant. For example, the sequence 4, 6, 8, 10, 12, 14, 16 is in arithmetic progression because the common difference between terms is 2. The terms of an arithmetic progression can increase or decrease infinitely depending on whether the constant difference between terms is positive or negative. Hence the partial arithmetic progression above can be partially written:

$$\dots - 10, \, - 8, \, - 6, \, - 4, \, - 2, \, 0, \, 2, \, 4, \, 6, \, 8, \, 10, \dots$$

The numerators and denominators of the sequence of ratios that appear to establish the proportions of bird's eggs as discussed previously are in arithmetic progression:

$$\frac{1}{2}; \ \frac{2}{3}; \ \frac{3}{4}; \ \frac{4}{5}; \ \frac{5}{6}; \ \frac{6}{7}; \ \frac{7}{8}; \ \frac{8}{9}; \ \frac{9}{10}; \ \frac{10}{11}; \ \dots$$

This progression of fractions contains an interesting attribute. With regard to any two adjacent fractions, if the numerators of either are multiplied by the denominators of the other, the products will always differ by one:

$$1 \times 3 = (2 \times 2) - 1$$

$$2 \times 4 = (3 \times 3) - 1$$

$$3 \times 5 = (4 \times 4) - 1$$

$$4 \times 6 = (5 \times 5) - 1$$

and so forth.

These equations can be written in another manner producing another fascinating array of equations:

$$3 = 2 + 2 - 1$$

$$4 + 4 = 3 + 3 + 3 - 1$$

$$5 + 5 + 5 = 4 + 4 + 4 + 4 - 1$$

$$6 + 6 + 6 + 6 = 5 + 5 + 5 + 5 + 5 - 1$$

$$7 + 7 + 7 + 7 + 7 + = 6 + 6 + 6 + 6 + 6 + 6 - 1$$

and so forth.

A **geometric progression** is defined as a sequence of numbers in which the ratio between any two successive terms is the same. For example, the progression 2, 4, 8, 16, 32, 64, 128, 256 is geometric because the ratio between each term is constant—in this case 2 as the terms increase in magnitude, or the reciprocal of two, ½, as the progression decreases in magnitude. The terms of a geometric progression can also increase or decrease infinitely:

$$\ldots, \ \frac{1}{32}; \ \frac{1}{16}; \ \frac{1}{8}; \ \frac{1}{4}; \ \frac{1}{2}; \ 1; \ 2; \ 4; \ 8; \ 16; \ 32; \ \ldots$$

A geometric progression can be written in terms of its constant ratio raised to a power, the exponents of which form an arithmetic progression (see Appendix C regarding the properties of exponents). For example, the geometric progression above can be written,

$$\ldots \ 2^{-5}; \ 2^{-4}; \ 2^{-3}; \ 2^{-2}; \ 2^{-1}; \ 2^{0}; \ 2^{1}; \ 2^{2}; \ 2^{3}; \ 2^{4}; \ 2^{5}; \ \ldots$$

It will be shown that this relationship—a series of exponents in arithmetic progression corresponding to the terms of a geometric progression (whose values are determined by the exponents)—structures the graceful curve of the shell of the chambered nautilus.

The **harmonic series** is defined as the reciprocals of the terms of an arithmetic progression. For example, the reciprocals of the arithmetic progression 1, 2, 3, 4, 5, 6, create a harmonic series:

$$\frac{1}{1}; \ \frac{1}{2}; \ \frac{1}{3}; \ \frac{1}{4}; \ \frac{1}{5}; \ \frac{1}{6};$$

The reference to *harmonic* relates to music. Consider a length of thin wire fixed at each end and stretched until it is taut (such as a guitar string). If the wire is plucked it will vibrate producing a tone. If the length of wire is reduced according to a harmonic series, first by 1/2, then by 1/3, 1/4, 1/5, etcetera, different tones will be produced ranging from a lower to a higher frequency (a lower to a higher pitch). Various combinations of these tones can be arranged to produce harmonic music, which is to say, music that is pleasing to the ear.

The **Fibonacci sequence**, 0, 1, 1, 2, 3, 5, 8, 13, 21, 34, ... is neither an arithmetic progression (the difference between each term is not constant) or a geometric progression (the ratio between each term is not constant). The sequence is considered recursive, which means that each term is dependent on a relationship involving previous terms. Interestingly, the quotients of the ratios of successive terms do converge to a constant value, which to five decimal places is the irrational number 1.61803 known as the golden proportion:

$$0/_1 = 0$$

$$1/_1 = 1.0$$

$$2/_1 = 2.0$$

$$3/_2 = 1.50$$

$$5/_3 = 1.66...$$

$$8/_5 = 1.60$$

$$13/_8 = 1.625$$

$$21/_{13} = 1.615...$$

$$34/_{21} = 1.6195...$$

$$55/_{34} = 1.6176...$$

$$^{89}/_{55} = 1.61818...$$

$$^{144}/_{89} = 1.61798...$$

$$^{233}/_{144} = 1.61806...$$

$$^{377}/_{233} = 1.61803...$$

$$^{610}/_{377} = 1.618034...$$

$$^{987}/_{610} = 1.618033...$$

$$^{1597}/_{987} = 1.6180344...$$

$$^{2584}/_{1597} = 1.6180338...$$

$$^{4181}/_{2584} = 1.6180341...$$

$$^{6765}/_{4181} = 1.61803396...$$

$$^{10946}/_{6765} = 1.61803399...$$

$$^{17711}/_{10946} = 1.618033985...$$

$$^{28657}/_{17711} = 1.618033990...$$

$$^{46368}/_{28657} = 1.618033988...$$

$$^{75025}/_{46368} = 1.618033989...$$

Interestingly, beginning with the ratio 2/1, the value of each successive ratio oscillates around the convergent value, either greater or less than (to five decimal places) 1.61803. For example, 1.61539 (21/13) is less than 1.61803; 1.61905 (34/21) is greater than 1.61803; 1.61765 (55/34) is less than 1.61803; 1.61818 (89/55) is greater than 1.61803; and so forth.

This oscillation relates to the product of the means and the extremes in an interesting manner. Considering any four successive terms of the Fibonacci sequence, the product of the means will always differ from the product of the extremes by the quantity of one. For example, considering the terms 13, 21, 34, 55 and 89:

$$21^2 = 441 = (13 \times 34) - 1$$

$$34^2 = 1156 = (21 \times 55) + 1$$

$$21 \times 34 = 714 = (13 \times 55) - 1$$

$$34 \times 55 = 1870 = (21 \times 89) + 1$$

and so forth. Thus, although the terms remain an additive recursive succession, the sequence tends toward a geometric progression where the product of the means equals the product of the extremes.

Another example of the occurrence of Fibonacci ratios can be found in the infinitely continued fraction,

$$1 + \cfrac{1}{1 + \cfrac{1}{1 + \cfrac{1}{1 + \cfrac{1}{1 + \cfrac{1}{1 + \cfrac{1}{1 + \cfrac{1}{1 + \cfrac{1}{1 + \cfrac{1}{1 + \cfrac{1}{1 \ldots}}}}}}}}}}$$

The successive terms of this infinitely continued fraction are successive Fibonacci ratios. The first term is

$$1 + \frac{1}{1} = 2$$

The next term is:

$$1 + \cfrac{1}{1 + \cfrac{1}{1}} = 1 + \frac{1}{2} = \frac{3}{2}$$

The next term is:

$$1 + \cfrac{1}{1 + \cfrac{1}{1 + \cfrac{1}{1}}} = 1 + \cfrac{1}{\cfrac{3}{2}} = 1 + \cfrac{2}{3} = \cfrac{5}{3}$$

The next term is:

$$1 + \cfrac{1}{1 + \cfrac{1}{1 + \cfrac{1}{1 + \cfrac{1}{1}}}} = 1 + \cfrac{1}{\cfrac{5}{3}} = 1 + \cfrac{3}{5} = \cfrac{8}{5}$$

Continuing in a like manner produces a series of Fibonacci ratios whose numerators and denominators correspond to the terms of the Fibonacci sequence,

$$\frac{2}{1}; \quad \frac{3}{2}; \quad \frac{5}{3}; \quad \frac{8}{5}; \quad \frac{13}{8}; \quad \frac{21}{13}; \quad \frac{34}{21}; \quad \ldots$$

The 23th iteration of the fraction (75025/46368) produces (to nine decimal places) the irrational number 1.618033989:

$$1.618033989 = 1 + \cfrac{1}{1 + \cfrac{1}{1 + \cfrac{1}{1 + \cfrac{1}{1 + \cdots}}}}$$

The Golden Proportion

Remarkably, 1.618033989... is also the ratio of the famous golden proportion, also referred to as the golden mean. Because of the profundity of this proportion with regard to the mathematics of nature, it is also called the Divine proportion. Even more remarkable is that it is derived from proportional ratios relating two segments— b and a—of a line:

Length b divided by length a is equal to the total length $(b + a)$ divided by length b:

$$\frac{b}{a} = \frac{b + a}{b}$$

Multiplying the means by the extremes results in a quadratic equation:

$$b^2 = ab + a^2$$

$$b^2 - ab - a^2 = 0$$

If $a = 1$, $\qquad b^2 - b - 1 = 0$

The solution for b (see appendix D) is,

$$b = \frac{1 \pm \sqrt{5}}{2}$$

resulting in $b = +1.6180339$ or -0.618033989 (to nine decimal places). Because b/a is greater than one, the positive number, $+1.61803389$ is taken as the golden proportion.

In the early part of the 20th century, the Greek letter ϕ (phi, pronounced phEYE) was adopted to designate the proportion.[1] Phi is the first letter of the 5th century Greek sculptor Phidias's name (thought to be one of the greater sculptors of ancient Greece, whose works were displayed in the Parthenon).

Substituting $b = \phi$ in the equation $b^2 - b - 1 = 0$ results in:

$$\phi^2 - \phi - 1 = 0$$

And therefore: $$\boldsymbol{\phi^2 = \phi + 1}$$

Multiplying both sides of this equation by $1/\phi$ results in:

$$\phi = 1 + 1/\phi.$$

Hence,

$$\frac{1}{\phi} = \phi - 1$$

Thus phi has two unique properties: (1) phi squared equals phi plus one, and (2) the reciprocal of phi equals phi minus one.

Historians believe that the golden proportion was discovered by the Egyptians and later introduced to Greek culture. An analysis of the measurements of the great pyramid at Giza constructed during the reign of Cheops (pronounced KEE-ops) demonstrates that indeed, the Egyptians were aware of ϕ.

With regard to the following drawing, the average length of each base b measures 230.36 meters.[2] The angle α of the slope of each face measures 51 degrees, 50 minutes and 40 seconds (51.84°).[3] Therefore, since the tangent of α is equal to the height h divided by one half the base b, h is equal to $1/2 \times 230.36$ meters $\times$

---

[1] H.E. Huntly, *The divine Proportion, (Dover, 1970) p.25*
[2] Peter Tomkins, *Secrets of the Great Pyramid* (Harper and Row, 1971) *p.366.*
[3] Tomkins, p.368

tan 51.84° = 146.58 meters. Hence the ratio of the height to the base is:

$$\frac{h}{b} = \frac{146.58}{230.36} = 0.636$$

Interestingly, $0.636 = \sqrt{\phi}/2$, hence, $\frac{h}{b} = \frac{\sqrt{\phi}}{2}$

And therefore,

$$h = \frac{b \times \sqrt{\phi}}{2}$$

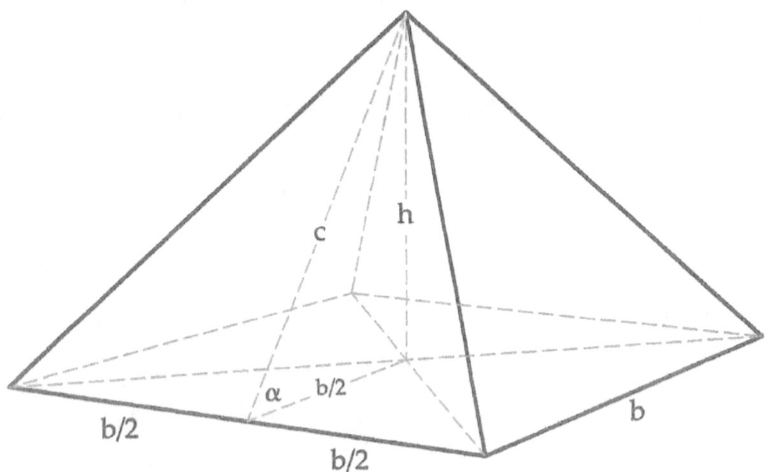

The Greek historian Herodotus stated that the area of each face of the pyramid is equal to the square of the height.[4] With regard to the drawing above, the area A of one face is equal to one-half the base b times the height c. The Pythagorean theorem can be used to determine c, knowing the value of height h and base b.[5]

---

[4] Tomkins, p.370
[5] The square of the hypotenuse of a right triangle is equal to the sum of the squares of the opposite sides.

With regard to the drawing of the pyramid on the preceding page:

$$c^2 = h^2 + \left(\frac{b}{2}\right)^2$$

Substituting $h = \frac{b\sqrt{\phi}}{2}$ and $b = 2$, $h = \sqrt{\phi}$. Therefore:

$$c^2 = \sqrt{\phi^2} + 1^2$$

$$c^2 = \phi + 1$$

Since $\phi^2 = \phi + 1$, $\qquad$ $c^2 = \phi^2$

and therefore, $\qquad$ $c = \sqrt{\phi^2} = \phi$

Since area A of each face equals $\frac{b}{2} \times c$, substituting $b = 2$ and $c = \phi$ results in:

$$A = \phi$$

Since height h is equal to $\sqrt{\phi}$, area A resulting from the squaring of the height is:

$$A = \sqrt{\phi^2}$$

$$A = \phi$$

The value of c in meters is derived using either the Pythagorean theorem or proportional ratios. Regarding proportional ratios:

$$\frac{c \ (in \ meters)}{230.36 \ m} = \frac{\phi}{2}$$

$$c = \frac{\phi \ (230.36)}{2}$$

$$c = 186.365 \text{ meters}$$

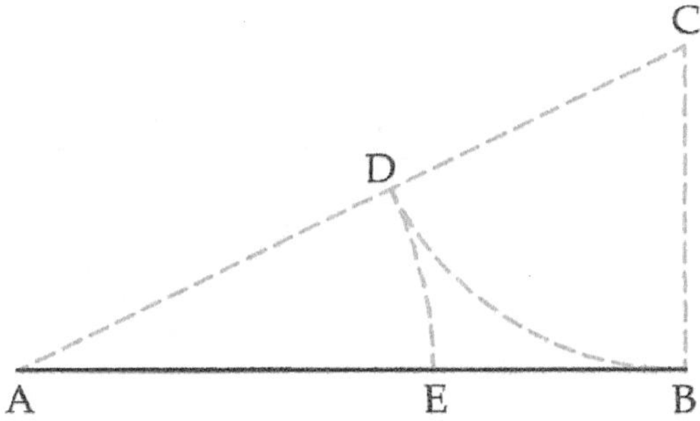

The division of any line of length *AB* into two segments *AE* and *EB* whose ratio is the golden proportion (*AE/EB* =ϕ), can be drawn geometrically as follows. Line *CB* = *AB*/2 is drawn perpendicular to *AB* at *B*. Line *AC* is drawn. Using *BC* as a radius, an arc is drawn intersecting *AC* at *D*. Using *AD* as a radius, an arc is drawn intersecting *AB* at *E*, resulting in *AE:EB* = ϕ. (See appendix E for the proof.)

The golden rectangle whose sides are proportioned according to the phi ratio (long side is to short side as phi is to one) is constructed from a square as follows:

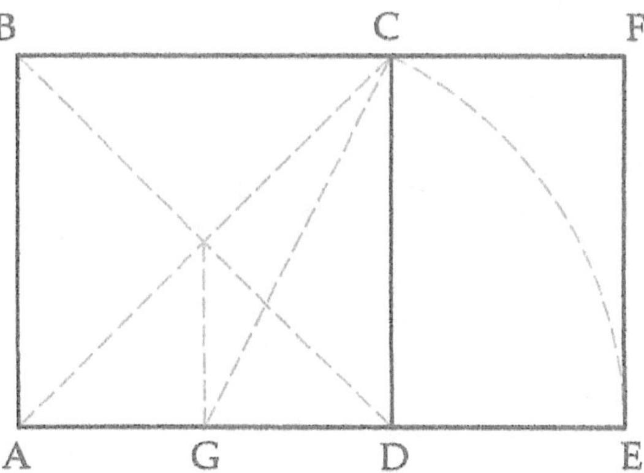

The midpoint of line *AD* is *G*. Using *GC* as a radius, an arc is drawn intersecting the extension of line *AD* at *E*. If *AD* is equal to one, *AE* = ϕ and *DE* = $\phi - 1 = 1/\phi$. (See appendix F for the proof.)

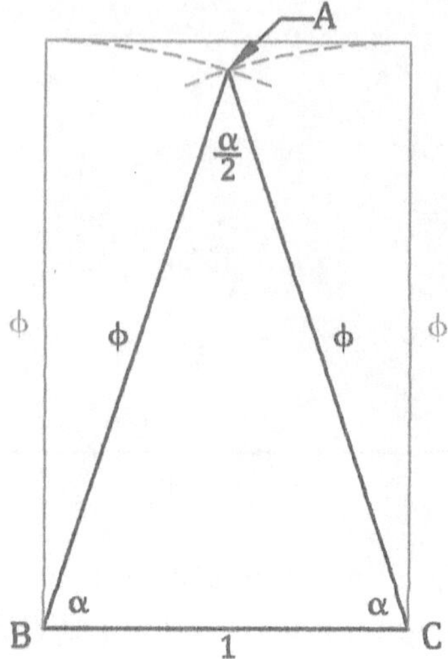

The golden triangle constructed from a golden rectangle (both with sides equal to phi and base equal to one, as shown above) is an isosceles triangle whose apex angle is one-half of either base angle α. The only base angle that satisfies this criterion is 72°.

$$\frac{\alpha}{2} + \alpha + \alpha = 180°$$

$$\frac{5\alpha}{2} = 180°$$

$$\alpha = 72°$$

The relationship $AB:BC = \phi$ can be substantiated using the proportional ratios of the law of sines (see appendix G):

$$\frac{AB}{\sin 72°} = \frac{BC}{\sin 36°}$$

$$\frac{AB}{BC} = \frac{\sin 72°}{\sin 36°}$$

$$\frac{AB}{BC} = 1.618033989 \ldots$$

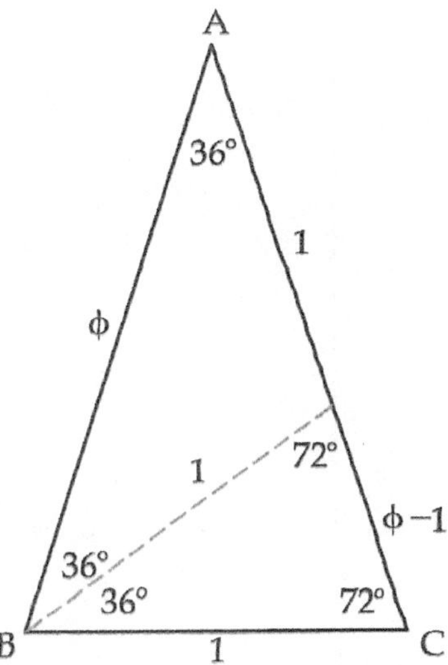

The value of phi as related to the diagram shown above can also be proved algebraically using the proportional ratios of similar triangles to derive the quadratic equation defining phi:

$$\frac{\phi}{1} = \frac{1}{\phi - 1}$$

$$\phi^2 - \phi - 1 = 0$$

$$\phi = \frac{1 + \sqrt{5}}{2}$$

$$= 1.618033989 \ldots$$

The familiar 3-4-5 triangle—where $3^2 + 4^2 = 5^2$—shares a unique relationship with phi and a circle (see diagram next page). The bisector of angle BCA meets AB at F. The circle is scribed with center F and radius FB, meeting AC at tangent point D. Line CF intersects BD at G, and the circle at H. The extension of CF intersects the circle at E. Using the Pythagorean theorem and the proportional relationships between similar triangles, the following Fibonacci and phi proportions can be derived:

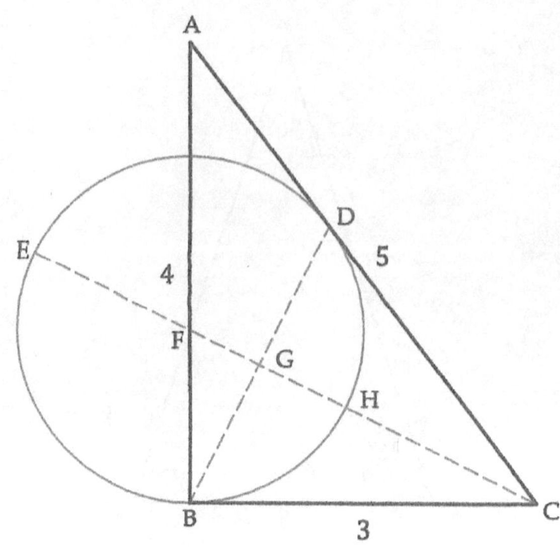

$$\frac{BC}{BF} = \frac{3}{3/2} = 2$$

$$\frac{DC}{AD} = \frac{3}{2}$$

$$\frac{AF}{FB} = \frac{5/2}{3/2} = \frac{5}{3}$$

$$\frac{AB}{FB} = \frac{8/2}{3/2} = \frac{8}{3}$$

$$\frac{EC}{EH} = \frac{3(1+\sqrt{5})/2}{3} = \phi$$

$$\frac{EH}{HC} = \frac{3}{3/2\left(\sqrt{5}-1\right)} = \phi$$

$$\frac{FG}{GH} = \frac{(3/10)\sqrt{5}}{(3/10)(\sqrt{5}-1)\sqrt{5}} = \frac{\phi}{2}$$

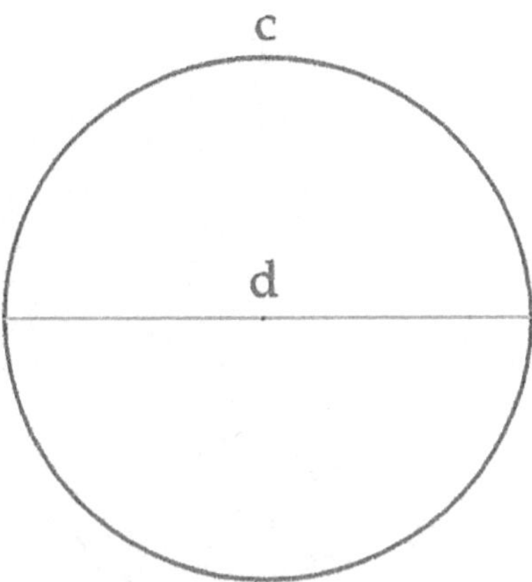

The value of the ratio relating the circumference c to the diameter d of a circle is 3.14159 (to five decimal places), and is designated by the Greek letter π:

$$\frac{c}{d} = \pi$$

The ratio of phi squared to pi is surprisingly close to the ratio of five to six:

$$\frac{\phi^2}{\pi} = 0.833346101 \dots$$

$$\frac{5}{6} = 0.833333333 \dots$$

* * * * * * * * * * * * * * * * *

The pentagram star (see next page) is made up of golden triangles and is therefore, a rich source of golden ratios. In the diagram below, R (*PA, PB, PC etc.*) is the radius of the larger circle; and r (*PI, PJ, PF, etc.*) is the radius of the smaller circle. If *HI* is equal to one, *AH* equals phi and the following relationships hold true:

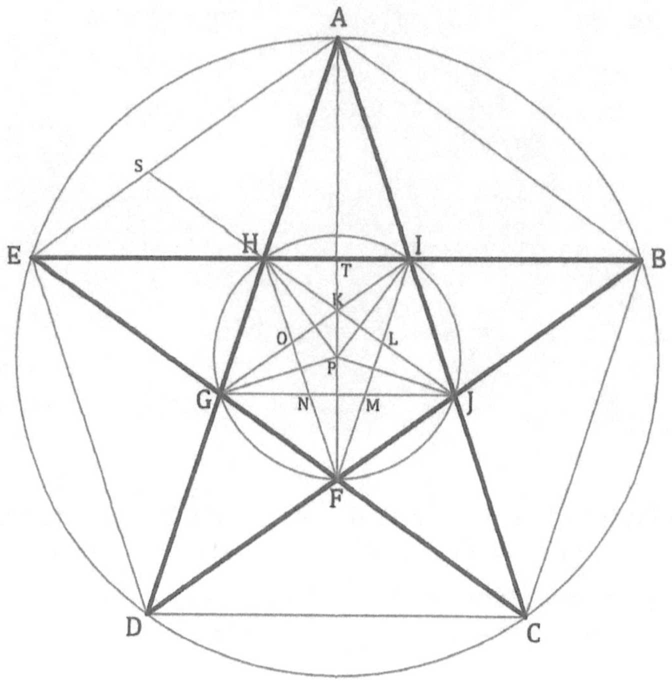

Radius *r* = *PF* = *PG* = *PH* = *PI* = *PJ*; radius **R** = *PA* = *PB* = *PC* = *PD* = *PE*.
See appendix *P* for the construction of a pentagon and pentagram star.

$$IF = FH = HJ = JG = GI = \phi$$

$$AB = BC = CD = DE = \phi + 1 = \phi^2$$

$$\frac{PT}{r} = \frac{\phi}{2}$$

$$\frac{R}{r} = \phi^2$$

$$\frac{PA}{PT} = 2\phi$$

$$\frac{FO}{OH} = \frac{IO}{OG} = \frac{EO}{OJ} = \frac{AS}{SE} = \frac{EH}{HI} = \frac{EI}{IB} = \phi$$

The following lengths associated with the pentagram are in geometric progression:

$$EB = \phi^3$$

$$EI = \phi^2$$

$$EH = \phi^1 = \phi$$

$$HI = \phi^0 = 1$$

$$HO = \phi^{-1}$$

$$ON = \phi^{-2}$$

The pentagonal shape occurs frequently in natural forms. One of its more popular appearances is in the cross section of the core of an apple:

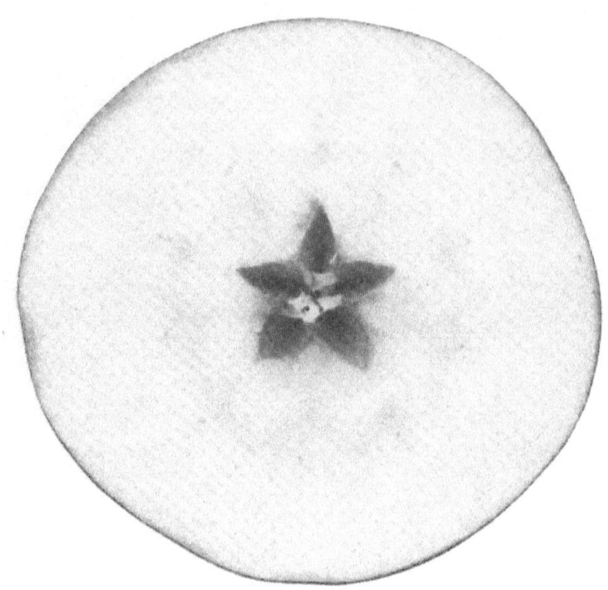

Apple blossoms always have five petals.

The nodules on the shell of a sea urchin (shown with spines removed)
are arranged according to two groups of radial pentagonal patterns.

The petals of numerous flowers are arranged according to the pentagon.

Fractals

The arrangement of the florets on a head of cauliflower according to parastiches reveals another interesting aspect of many of nature's shapes and forms: self-similarity. With regard to the photograph above, one of the clusters of florets that had been attached to the main stalk is shown removed, and placed to the left of the head (which is shown on the right). Note that the removed cluster resembles the original head, but at a reduced scale. The next smallest floret (shown on the far left) has been removed from the first cluster. It too resembles the original cauliflower head but at a reduced scale. In a fascinating manner, a head of cauliflower is made up of portions which resemble the whole.

This phenomenon of self-similarity is associated with fractal geometry. The word fractal (from the Latin *fractus* meaning broken or fractured) was coined by mathematician Benoit Mandelbrot in 1975 when the discipline of mathematics called chaos theory was being discovered and formulated by computer scientists. Mandelbrot needed a word to describe the geometry of natural forms whose component parts resembled a reduced or fragmented version of the whole.

Fractal geometry involves geometric progressions. When a shape or form is made up of components which themselves are fractions (ratios) of the original shape or form, progressions and infinite series appear. Consider for example, the transformation of an equilateral triangle,

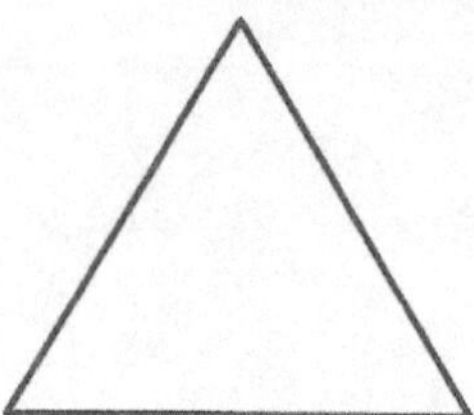

by means of the continued addition of equilateral triangles that are a fraction of the original size. For example, three triangles whose sides are one-third the length of the original are added to each of the sides of the original triangle:

Continuing in a like manner, additional triangles whose sides are one-third the length of the sides of the previously added triangles are added to the sides of the previously added triangles:

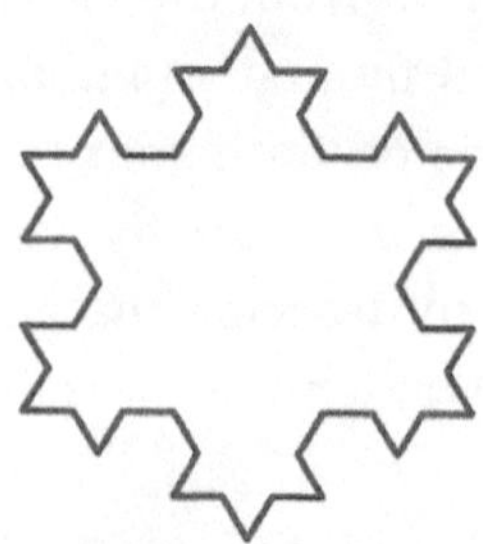

A fourth application of additional triangles transforms the previous configuration to the following configuration:

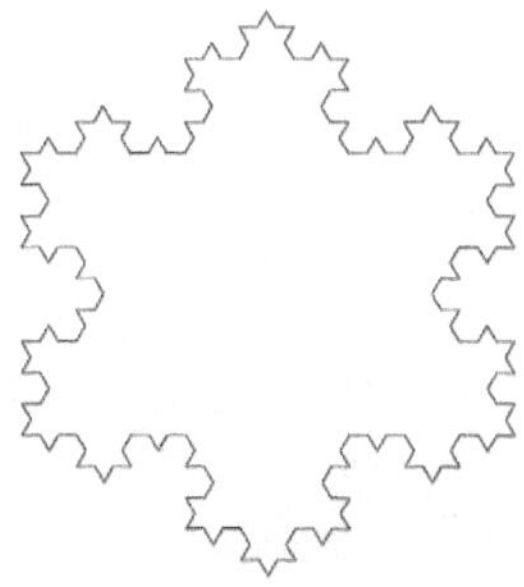

The transformation can continue again and again resulting in a figure called the Koch curve, named after the Swedish mathematician Helge Von Koch who first described it in 1904. The essence of the curve begins to become established after about the fifth transformation:

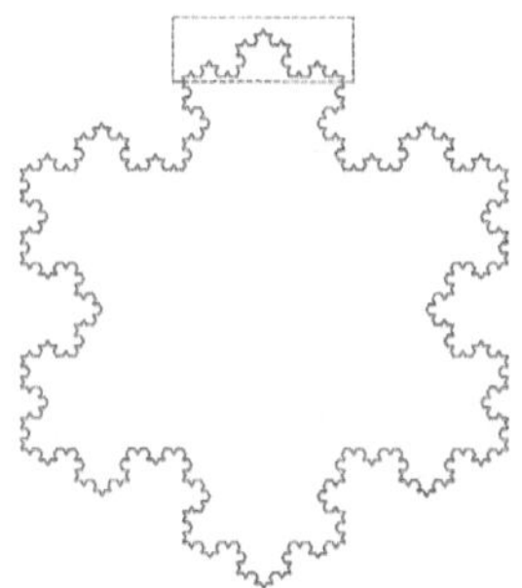

If the upper portion of the fifth transformation (shown in the dashed rectangle above) were to be enlarged, it would appear as shown below:

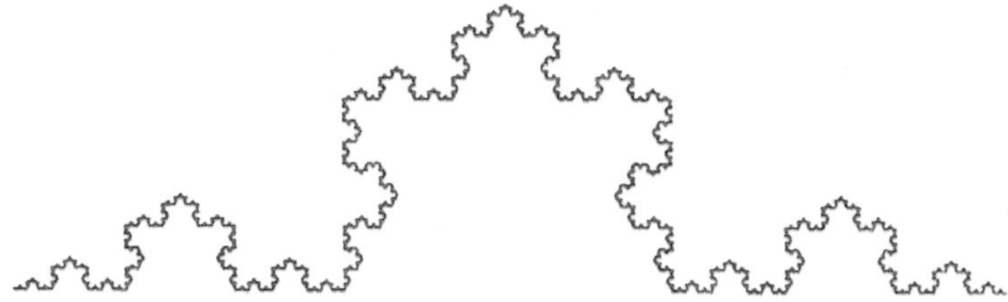

If, after five more transformation, the corresponding upper portion of the enlargement were to be enlarged, it would look exactly like the enlargement itself. The transformations can continue indefinitely. If the length of the side of the original triangle is a, the infinitely long curve expands according to the geometric progression (see appendix S):

$$3a, \quad 3a\left(\frac{4}{3}\right), \quad 3a\left(\frac{4}{3}\right)^2, \quad 3a\left(\frac{4}{3}\right)^3, \quad 3a\left(\frac{4}{3}\right)^4 \ldots$$

For example, if $a = 1$, the perimeter of the starting triangle is 3. The perimeter of the next (first) transformation is 4; of the second transformation, $3\times(4/3)^2 = 5.333\ldots$; of the third, $3\times(4/3)^3 = 7.111\ldots$; and so forth. It is somewhat difficult to imagine that if $a =$ one inch, the length of the curve after the fiftieth iteration is $3\times(4/3)^{50} = 5,297,343$ inches or 83.6 miles! It is even more difficult to imagine that the length of the curve is infinite when, paradoxically, the area contained by the curve is finite!

The area (A_n) subtended by the curve after an infinite number of iterations can be determined from an equation factoring the area of the original triangle (A_0) with an infinite series (see appendix T):

$$A_n = A_0 \times \left[1 + 3\left(\frac{1}{9} + \frac{4}{9^2} + \frac{4^2}{9^3} + \frac{4^3}{9^4} + \ldots\right)\right]$$

$$A_n = A_0 \times 1.6 = A_0 \times 8/5$$

$$\frac{A_n}{A_0} = \frac{8}{5}$$

Remarkably, the area generated by the Koch curve is related to the area of the original triangle according to the Fibonacci ratio 8/5!

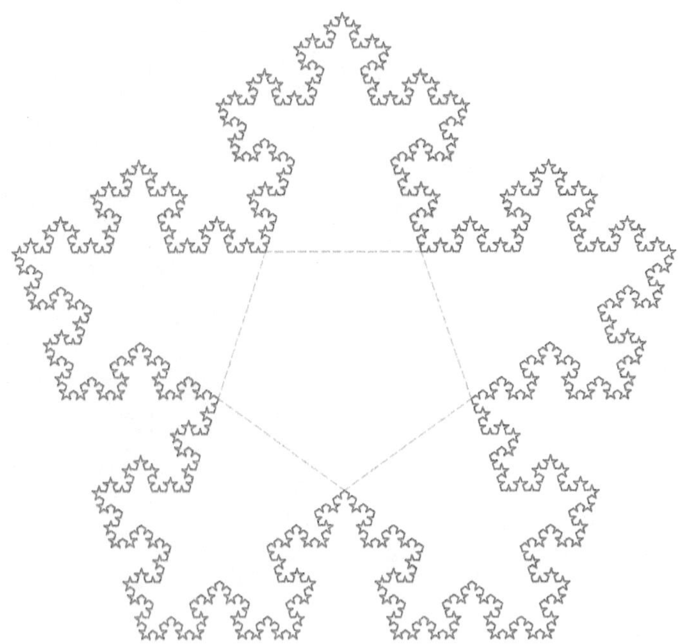

Fourth iteration of a pentagram Koch-curve.

The Koch-type curve generated from a pentagram (made up of five golden triangles, with sides ϕ and base 1, attached to the sides of a pentagon) expands in length according to the geometric progression (see appendix W),

$$10\phi, \;\; 10\phi\left(\frac{4}{\phi^2}\right), \;\; 10\phi\left(\frac{4}{\phi^2}\right)^2, \;\; 10\phi\left(\frac{4}{\phi^2}\right)^3, \;\; 10\phi\left(\frac{4}{\phi^2}\right)^4, \ldots$$

The area after an infinite number of iterations is:

$$A_{n\to\infty} = A_0 + \frac{1}{2}\left(5\sqrt{\phi + 3/4} + \frac{10\sqrt{\phi + 3/4}}{\phi^4 - 4}\right)$$

where A_0 is the area of the initial pentagon. Thus, although the length of the curve is infinite, the area after an infinite number of iterations is 8.2635, corresponding to the area of the original pentagon equal to 5.5676. (See appendix X.)

Golden Spirals

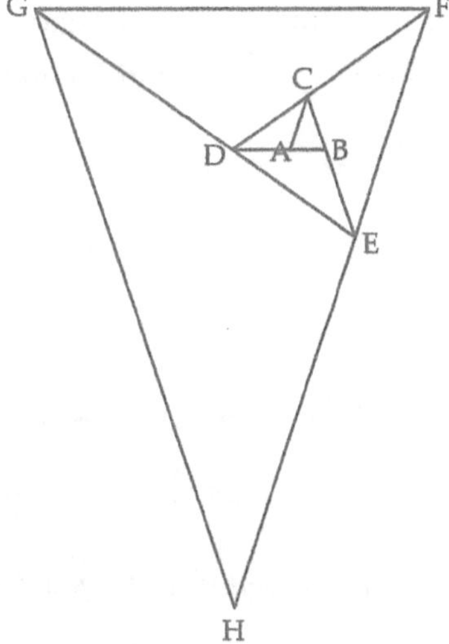

The diagram above illustrates a series of golden triangles where the side of one triangle is also the base of another. For example, EF is the base of triangle EGF and also a side of triangle DFE. If $AB = 1$, and $BC = \phi$, the sides of each successive larger triangle expand according to a geometric progression corresponding to a Fibonacci sequence:

$$AB = \phi^0 = 1$$

$$BC = \phi^1 = \phi$$

$$CD = \phi^2 = \phi + 1$$

$$DE = \phi^3 = 2\phi + 1$$

$$DE = \phi^4 = 3\phi + 2$$

$$EF = \phi^5 = 5\phi + 3$$

$$FG = \phi^6 = 8\phi + 5$$

The geometric progression and Fibonacci sequence are derived knowing that each triangle is proportional to the other, and that $\phi^2 = \phi + 1$. Additionally, the area of each successive triangle also expands according to a geometric progression and a sequence of alternating Fibonacci terms. If the area of triangle ABC is unity,

$$ABC = \phi^0 = 1$$

$$BCD = \phi^2 = \phi + 1$$

$$CDE = \phi^4 = 3\phi + 2$$

$$DEF = \phi^6 = 8\phi + 5$$

$$EFG = \phi^8 = 21\phi + 13$$

$$FGH = \phi^{10} = 55\phi + 34$$

Connecting similar points on the rotating golden triangles with a smooth curve produces a golden triangle logarithmic spiral:

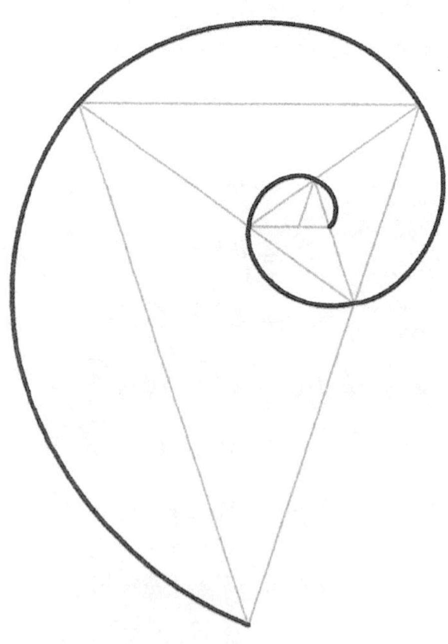

Connecting similar points on the rotating golden rectangles (or similar points on their associated rotating squares) with a smooth curve produces a golden rectangle logarithmic spiral:

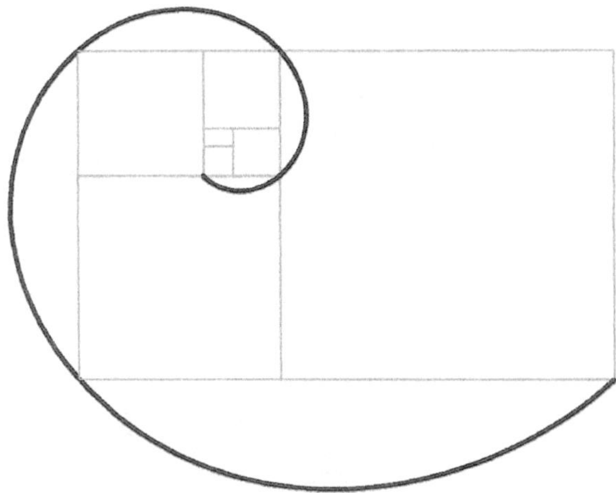

The logarithmic spiral generated by the golden rectangle is not the same as the spiral generated by the golden triangle. If the two spirals (shown below) share the same pole point P and starting point A, the curves are different. (See Appendices M and N.)

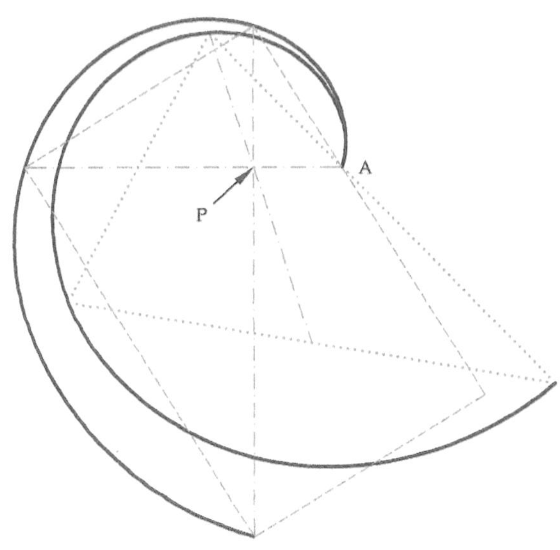

The Logarithmic Spiral

Understanding the logarithmic spiral begins with an understanding of the special relationship between the curve of the spiral, and arithmetic and geometric progressions. A spiral curve is considered logarithmic if each point of the curve represents the term of an arithmetic progression corresponding to a term of a geometric progression. The point on the curve representing the term of the arithmetic progression is the logarithm of the corresponding term of the geometric progression. The logic behind this relationship begins with a review of the definition of a logarithm.

A logarithm is the exponent expressing the power to which a fixed number (the base) must be raised in order to produce a given number (the antilogarithm). Using letters to represent numbers, this definition is expressed by the equation,

$$a = n^s$$

The exponent s (expressing the power to which n must be raised to produce a) is the logarithm to the base n of a. This is written," $\log_n a = s$," and is read "the logarithm of a to the base n, is s," or, conversely, "s is the logarithm to the base n of a."

The relationship between logarithms and arithmetic and geometric progressions is best explained with an example. Consider a sequence of terms established by a base number, say 2, whose values increase or decrease according to an arithmetic progression of exponents (whose common difference in this example is one):

$$2^0 \quad 2^1 \quad 2^2 \quad 2^3 \quad 2^4$$

The terms of this sequence correspond to the terms of a geometric progression (whose common ratio between increasing terms is 2, and decreasing terms is 1/2):

$$1 \quad 2 \quad 4 \quad 8 \quad 16$$

The exponents 0, 1, 2, 3, and 4 forming the arithmetic progression are the logarithms to the base 2 of the terms of the corresponding geometric progression: $\log_2 1 = 0$, $\log_2 2 = 1$, $\log_2 4 = 2$, $\log_2 8 = 3$, and so forth. This corresponding relationship between the two progressions can be expressed graphically using lines emanating from a central point P:

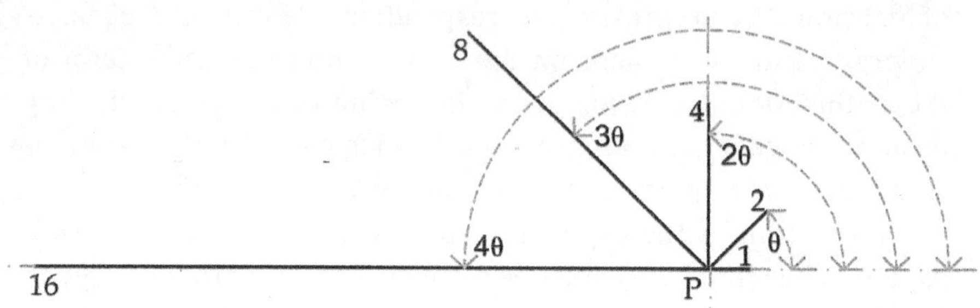

The lengths of the lines (1, 2, 4, 8, and 16) correspond to the magnitudes of the terms of the geometric progression. The directions of the lines are established by angles (θ, 2θ, etc.) measured from a base line. Since the lines have both magnitude and direction, and emanate from the same central point P, they are called radius vectors and are designated by the letter r ($r = 1, 2, 4$, etc).

The angles correspond to the respective terms of the arithmetic progression. Hence the angular displacements (θ, 2θ, 3θ and 4θ) represent the logarithms (to the base 2) of the terms of the geometric progression: $\log_2 r = b\theta$, (where r equals any radius vector).

Additional terms of the corresponding progressions,

2^{-4}	2^{-3}	2^{-2}	2^{-1}	2^0	2^1	2^2	2^3	2^4	2^5	2^6	2^7
$\frac{1}{16}$	$\frac{1}{8}$	$\frac{1}{4}$	$\frac{1}{2}$	1	2	4	8	16	32	64	128

can be added to the graph in a similar manner such that the exponents in arithmetic progression continue to increase in a counterclockwise direction (along with the magnitude of the terms of the geometric progression) and decrease in a clockwise direction:

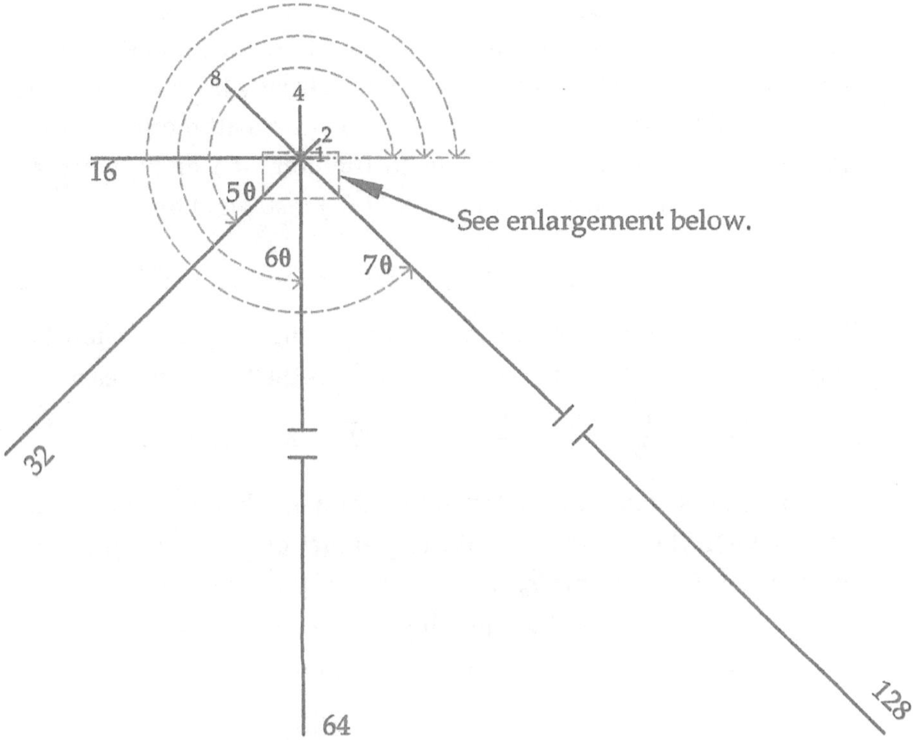

See enlargement below.

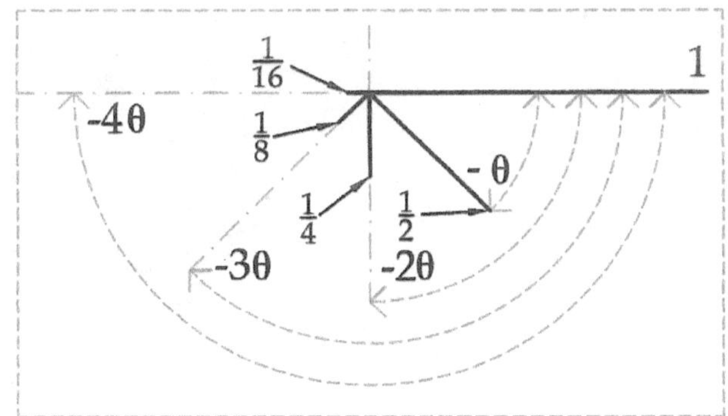

The radius vectors can be plotted over or under each other as they increase or decrease in magnitude: e.g. 128 is directly above 1/2; 1/4 is directly below 64, and so forth.

Additional radius vectors (indicated by the dashed lines in the following diagrams) can be added to the graph, positioned equidistant between the radius vectors that have already been plotted. The common difference between the terms of the arithmetic progression of exponents (corresponding to the common angle between the radius vectors) becomes one-half (1/2) instead of one:

$$... \ \mathbf{2^{-2.5}} \quad \mathbf{2^{-2}} \quad \mathbf{2^{-1.5}} \quad 2^{-1} \quad \mathbf{2^{-0.5}} \quad 2^{0} \quad \mathbf{2^{0.5}} \quad 2^{1} \quad \mathbf{2^{1.5}} \quad 2^{2} \quad \mathbf{2^{2.5}} \ ...$$

The common ratio of the terms of the geometric progression becomes $\sqrt{2}$ as the terms increase, and $1/\sqrt{2}$ as the terms decrease:

$$... \ \frac{1}{4\sqrt{2}} \quad \frac{1}{4} \quad \frac{1}{2\sqrt{2}} \quad \frac{1}{2} \quad \frac{1}{\sqrt{2}} \quad 1 \quad \sqrt{2} \quad 2 \quad 2\sqrt{2} \quad 4 \quad 4\sqrt{2} \ ...$$

These new values are determined knowing that any radius vector r is equal the square root of the product of the adjacent radius vectors, say for example r_a and r_b: $r^2 = r_a \times r_b$ and therefore, $r = \sqrt{r_a \times r_b}$. With regard to the following two diagrams, the numbers in parentheses represent the angles that determine the direction of each radius vector:

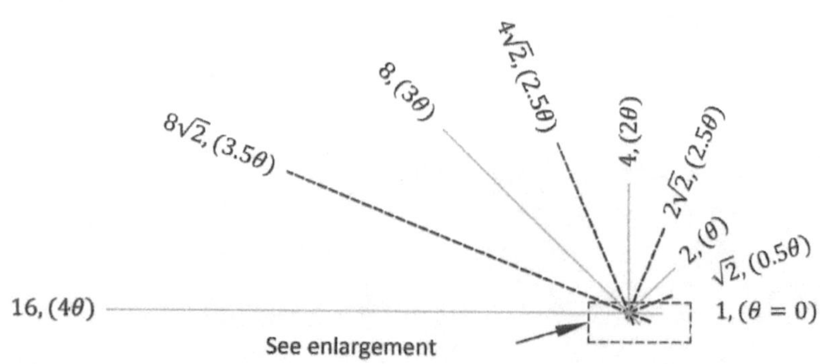

Hence for example, the magnitude of the radius vector located at angle (3.5θ) is:

$$r_{(3.5\theta)} = \sqrt{r_{(3\theta)} \times r_{(4\theta)}}$$

$$= \sqrt{8 \times 16} = 4\sqrt{8} = 4\sqrt{2 \times 4} = 8\sqrt{2}$$

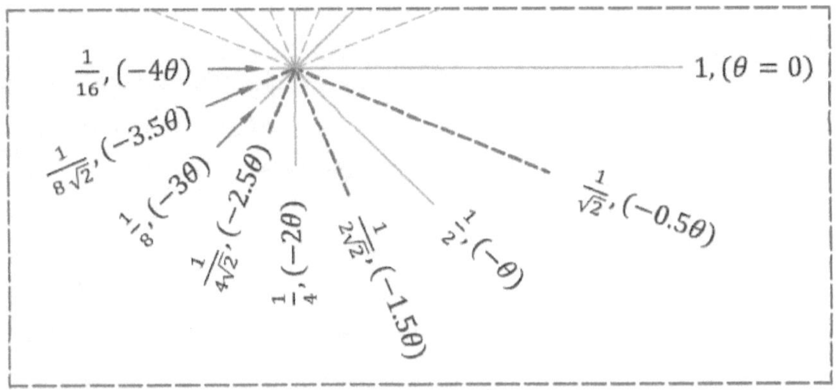

The magnitude of each new decreasing radius vector can also be determined knowing that the ratio between any radius vector on one side of the pole, and the radius vector on the opposite side of the pole (on the same vector line), is the same. In this example, the ratio is 16:1 or conversely, 1:16. (Note that with regard to this example, the proportional ratio $\sqrt{2}/2 = 1/\sqrt{2}$ can be used to simplify the values of r where applicable.)

$$r_{(-0.5\theta)} = \frac{1}{16} \times r_{(3.5\theta)} = \frac{1}{16} \times 8\sqrt{2} = \frac{\sqrt{2}}{2} = \frac{1}{\sqrt{2}}$$

$$r_{(3.5\theta)} = 16 \times r_{(-0.5\theta)} = 16 \times \frac{1}{\sqrt{2}} = 16 \times \frac{\sqrt{2}}{2} = 8\sqrt{2}$$

Regarding the previous diagrams, it can be seen that the magnitudes of the decreasing radius vectors below the base line are also the reciprocals of the magnitudes of the corresponding, increasing radius vectors above the base line. For example, $1/\sqrt{2}$ at the angle located and designated by (-0.5θ) is the reciprocal of $\sqrt{2}$ at (0.5θ); $1/2$ at $(-\theta)$ is the reciprocal of 2 at (θ); $1/8\sqrt{2}$ is the reciprocal of $8/\sqrt{2}$ and so forth.

Yet another set of radius vectors can be added to the graph in a like manner, each equidistant from the adjacent previously plotted radius vectors, using any of the aforementioned procedures: (1) the terms of a geometric progression corresponding to exponents of an arithmetic progression; (2) proportionality as related to the law of the means or (3) proportionality as related to reciprocals.

Regarding the terms of the geometric progression, the new common ratio of each term is $\sqrt{\sqrt{2}} = \sqrt[4]{2}$, (or its reciprocal, $1/\sqrt[4]{2}$) resulting in the values of the terms of a portion of the new geometric progression:

$$\ldots \frac{1}{\sqrt[4]{2}}, \quad 1, \quad \sqrt[4]{2}, \quad \sqrt{2}, \quad \sqrt{2}\sqrt[4]{2}, \quad 2, \quad 2\sqrt[4]{2}, \quad 2\sqrt{2}, \quad 2\sqrt{2}\sqrt[4]{2}, \quad 4, \ldots$$

Corresponding to (a portion) of the arithmetic progression:

$$\ldots \; 2^{-.25}, \quad 2^0, \quad 2^{.25}, \quad 2^{.5}, \quad 2^{.75}, \quad 2^1, \quad 2^{1.25}, \quad 2^{1.5}, \quad 2^{1.75}, \quad 2^2 \ldots$$

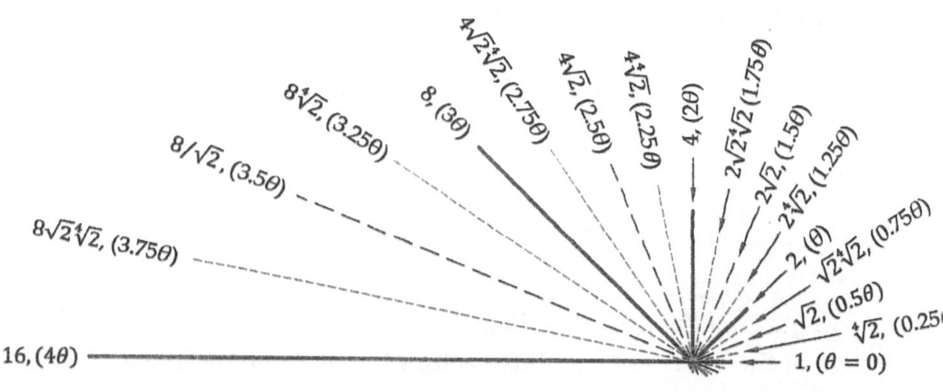

A continuous curved line can be drawn connecting the ends of the radius vectors:

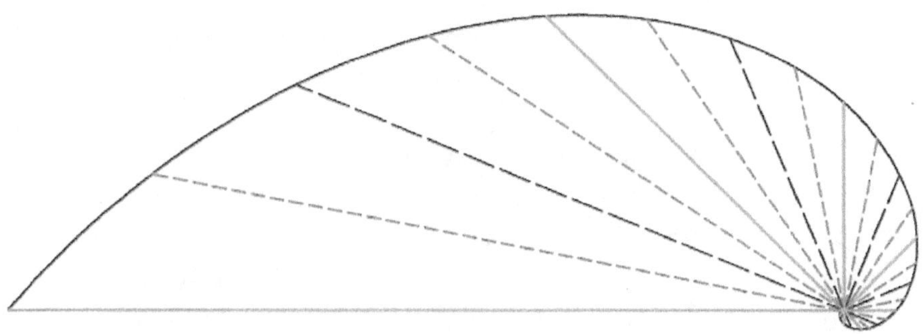

Since each point of the curve is established by an arithmetic progression, which is to say, the logarithm (establishing the direction and magnitude) of each radius vector, the curve can be said to be logarithmic.

This construction of a logarithmic spiral is in fact a graph expressed by a polar equation: a series of points (a curve) in a plane, established according to vectors originating from a central point (the pole) whose directions are established by angles measured from a horizontal base line. (See appendix H.)

It is apparent that the accuracy of the curve is dependent on the number of plotted radius vectors. The more radius vectors that are plotted, the more accurate the curve becomes. However, as demonstrated in the diagram of the curve above, accuracy can be achieved with only eight radius vectors for each quadrant of the curve.

The standard polar equation for any logarithmic spiral curve is,

$$r = ae^{b\theta}$$

The radius vector r represents the magnitude of any point of the curve with reference to the pole. The letter a determines the size or scale of the spiral, and e is a constant whose value is 2.71828... (see Appendix I). The exponent b determines the shape (pitch) of the spiral by factoring θ, the angle (measured in radians—see Appendix H) establishing the direction of each radius vector with relation to the horizontal axis and pole. Since $b\theta$ is the logarithm to the base e of r, it can also be said that the curve is logarithmic because its equation, $r = ae^{b\theta}$, is of the same type that defines a logarithm: $a = n^s$.

The value of b in the standard polar equation for the spiral constructed in the previous example (according to the exponentiation of the base number 2) is derived knowing $r = 16$ when $\theta = \pi$ radians (or 4 when $\theta = \pi/2$ radians, and so forth). If a is assigned the value of one, and log e is written ln, b is derived as follows:

$$r = ae^{b\theta}$$

$$16 = e^{b\pi}$$

$$ln\ 16 = b\pi$$

$$\frac{ln\ 16}{\pi} = b$$

$$r = e^{((ln16)/\pi)\theta}$$

$$r = e^{2.772759\theta/\pi} = e^{0.882540\theta}$$

The standard polar equation for the same spiral in terms of the base 2 is derived in a similar manner, also solving for b knowing that $r = 16$ when $\theta = \pi$:

$$16 = 2^{b\theta}$$

$$\log_2 16 = b\pi$$

$$\frac{\log_2 16}{\pi} = b = \frac{4}{\pi} = 1.27234$$

$$r = 2^{4\theta/\pi}$$

$$r = 2^{1.27320\theta}$$

In terms of the base 10 (where $\log_{10}$ is written log),

$$16 = 10^{b\pi}$$

$$log\,16 = b\pi$$

$$\frac{log\,16}{\pi} = b$$

$$r = 10^{[(log\,16)/\pi]\theta}$$

$$r = 10^{1.20412\pi/\theta}$$

$$r = 10^{0.38328\theta}$$

In terms of the base 16, $\qquad 16 = 16^{b\theta}$

$$log_{16}\,16 = b\pi$$

$$1 = b\pi$$

$$\frac{1}{\pi} = b$$

$$r = 16^{\theta/\pi}$$

It is apparent from these equations that the formula for the spiral can be expressed according to the exponentiation of any positive, real, base number. For example, the base number can represent a ratio, say k, such that $r = ak^{b\theta}$. If the ratio k represents two radius vectors, say r_1 and r_2 separated by π radians the polar equation for a logarithmic spiral, $r = ae^{b\theta}$, can be modified and written:

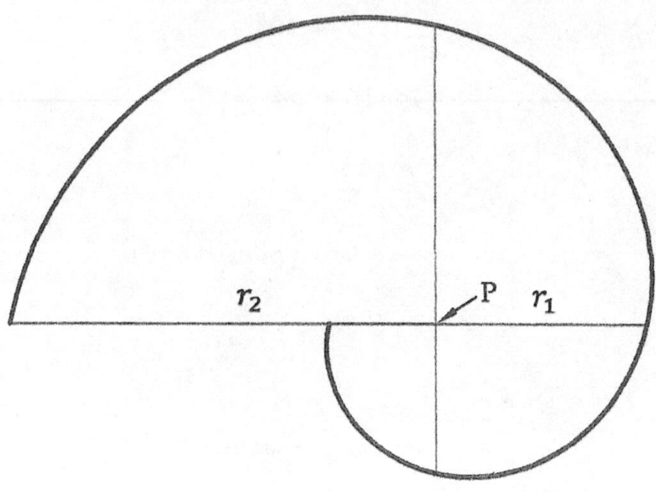

The derivation of $r = r_2^{\theta/\pi}$ is based on the standard equation $r = ae^{b\theta}$ where e is replaced with a constant k:

$$r = ak^{b\theta}$$

If $a = 1$,
$$r = k^{b\theta}$$

If $k = r_2/r_1$, and if $r_1 = 1$,
$$r = r_2^{b\theta}$$

If $\theta = \pi$,
$$r = r_2^{b\pi}$$

$$\log_{r_2} r_2 = b\pi$$

Since $\log_{r_2} r_2 = 1$,
$$1 = b\pi$$

$$\frac{1}{\pi} = b$$

Hence,
$$r = r_2^{\theta/\pi}$$

The equation, $r = r_2{}^{\theta/\pi}$, introduces a simple, yet systematic method for constructing a logarithmic spiral. The construction requires appropriate mechanical drawing equipment or appropriate computer software.

Step 1

Draw a series of equally spaced radius vectors. A convenient number is 16, producing the constant angle $2\pi/16$ radians (22.5°) between each radius vector. (Every other radius vector is shown dashed for clarity.)

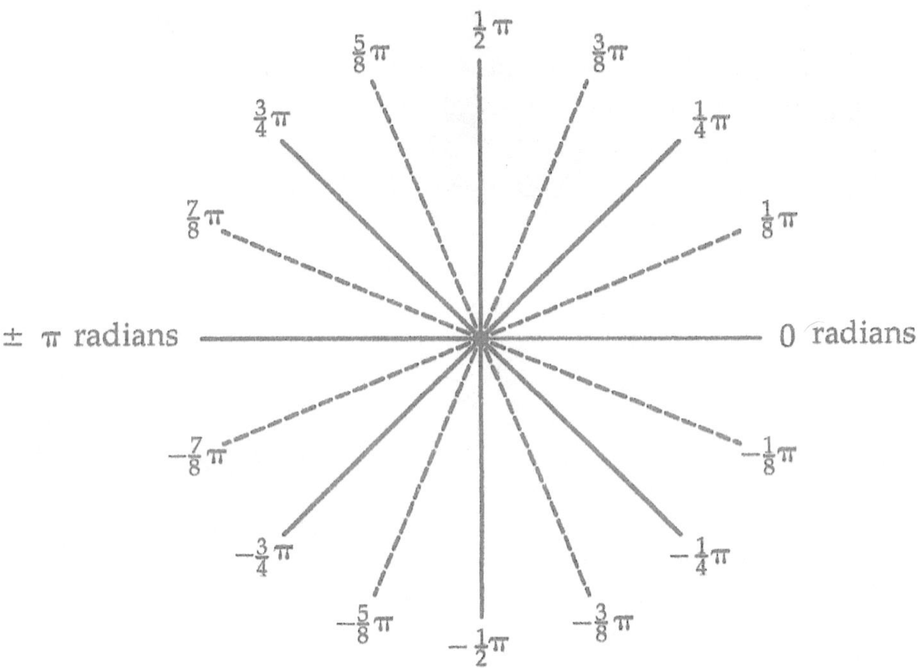

Step 2

Select a value for radius vector $r_2 = K$ (corresponding to $\theta = \pi$ radians). Establish radius vector $r_1 = K^0 = 1$ corresponding to $\theta = 0$ radians:

Step 3

Determine the values of the other radius vectors according to K raised to the exponents indicated by the fractions representing the arithmetic progression of the segments of the circle. Since π cancels out when substituted for θ in the equation $r = r_2^{\theta/\pi}$, (for example, $(\pi/8) \div \pi = 1/8$, $(2\pi/8) \div \pi = 1/4$, and so forth) the exponents become an arithmetic progression of eighth's: 1/8, 2/8, 3/8, 4/8 and so forth. (See drawing below.)

<u>Step 4</u>

Plot the radius vectors (corresponding to their respective exponents). (See also appendix R for alternate methods of determining the magnitude of the radius vectors.)

<u>Step 5</u>

Connect the ends of the radius vectors with a smooth curve (or to approximate the curve, use straight lines):

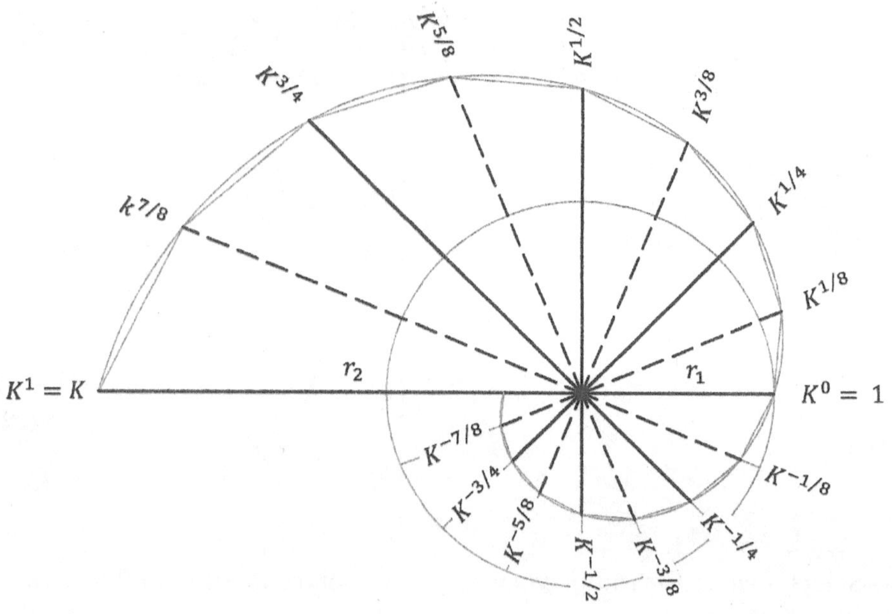

It is apparent from this construction, that the length of the spiral curve is equal to the sum of the lengths of the curve between the ends of each respective radius vector. Similarly the area defined by the curve is equal to the sum of the individual areas defined by the curve and the respective radius vectors. If the space between the radius vectors is made infinitely small, an accurate length and area of the spiral can be obtained.

Using calculus and the standard equation, $r = ae^{b\theta}$, the equation for the length L of the curve between any two radius vectors (see Appendix J) is:

$$L = \left(\frac{1}{b}\right)\left(\sqrt{1 + b^2}\right)(r_2 - r_1)$$

The equation for the area A between the curve and any two radius vectors positioned between r at $\theta = 0$ radians and r at $\theta = \pi$ radians (see Appendix K) is:

$$A = \left(\frac{1}{4b}\right)(r_2{}^2 - r_1{}^2)$$

The expansion and contraction of the spiral as it wraps around the pole occurs without changing the shape of the spiral. The shape remains the same as the size changes. For example, with reference to the horizontal base line and the vertical line through the pole as shown in the drawing below, (both of which divide the spiral into quadrants), the curve of the spiral shown dashed (where the value of r at $\theta = \pi$ radians is represented by K^{-1}) is the same shape as the spiral where $r = K^1$. (Although K can be given any positive value, in this example, K equals 2 resulting in the shape shown.)

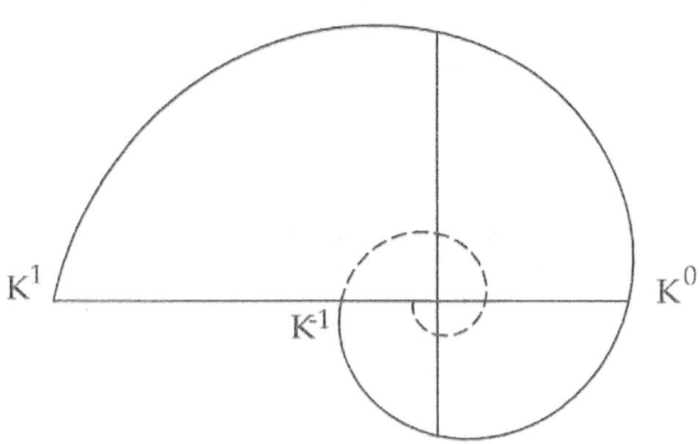

The length of the spiral shown dashed is $1/K^2$ times the original length (from the relationship $K^{-1}/K = 1/K^2$) Its area is $1/K^4$ times as large. If this smaller spiral is enlarged (such that K^{-1} is the same length as K^1) it will be exactly the same as the original spiral where r at π radians $= K^1$:

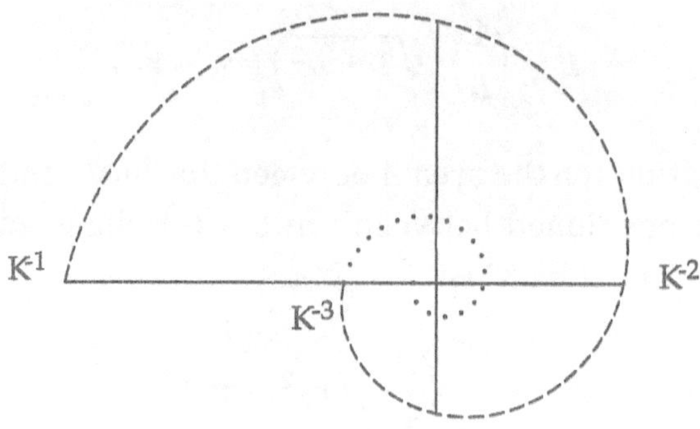

Likewise, if the smaller spiral represented by the dots (above) is enlarged so that K^{-3} is the same length as K^{-1} (below), the shape of the spiral represented by the dots will be the same as the spiral represented by the dashed line:

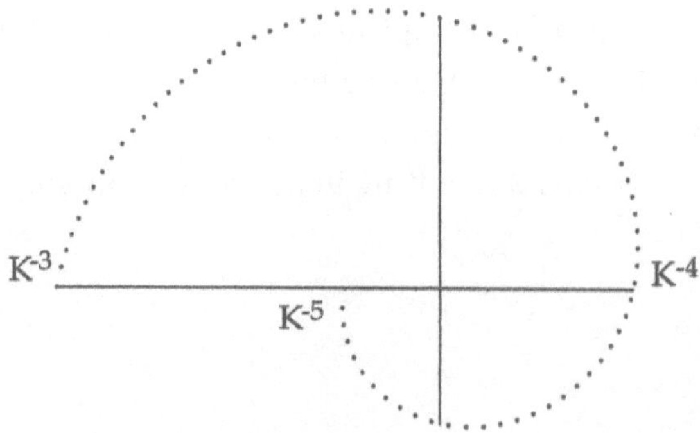

Substituting $b = lnK/\pi$ in the formula for the length of a spiral results in the equation:

$$L = \left(\frac{1}{(lnK)/\pi}\right)\left(\sqrt{1 + (lnK)/\pi}\right)(r_2 - r_1).$$

Using this formula, continuing with the substitution of 2 for K, *and* where $K^{-1} = r_1$ and $K^1 = r_2$, the length of the spiral from K^{-1} to K^1 is:

$$L = \left(\frac{1}{(.6932)/\pi}\right) \times \left(\sqrt{1 + (.6932)/\pi}\right) \times \left(2 - \frac{1}{2}\right)$$

$$L = (4.5331) \times (1.1048) \times (1.5) = 7.5123$$

Substituting for b in terms of K in the formula for the area A of a spiral results in the equation:

$$A = \left(\frac{1}{4(lnK)/\pi}\right)(r_2{}^2 - r_1{}^2)$$

Using this formula, continuing with the substitution of 2 for K, the area of the spiral from K^{-1} to K^1 is:

$$A = \left(\frac{1}{4(0.6932)/\pi}\right) \times \left(2^2 - \left(\frac{1}{2}\right)^2\right)$$

$$A = (1.1331) \times (3.75) = 4.2491$$

The length and area of the similar reduced spirals can be found in a similar manner. Note however, that the similar areas of the spirals shown dashed and dotted are contained in the area of the original spiral.

The constant shape of spiral is validated by the constant ratio between the values of K separated by π radians shown in the previous drawings:

$$\frac{K^{-4}}{K^{-3}} = \frac{K^{-2}}{K^{-1}} = \frac{K^0}{K^1} = K^{-1} = \frac{1}{K}$$

Another proof that the shape remains the same as the spiral expands or diminishes in size is the constant tangential angle α between the curve and any radius vector. With regard to the following drawing, it can be shown that $\tan \alpha = 1/b$. (See appendix L.)

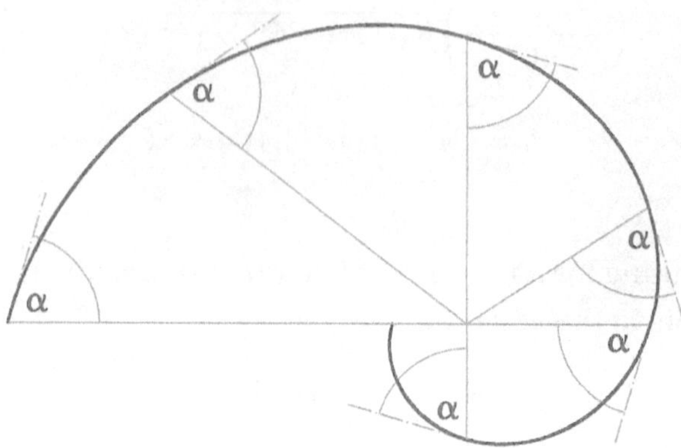

Because α is the same at any radius vector, the shape (also referred to as the pitch) of a logarithmic spiral is determined by the magnitude of α at the tangential slope. The greater α becomes, the "tighter" the spiral becomes. (Note that the angle determining the pitch of a spiral is equal to $90° - \alpha$.)

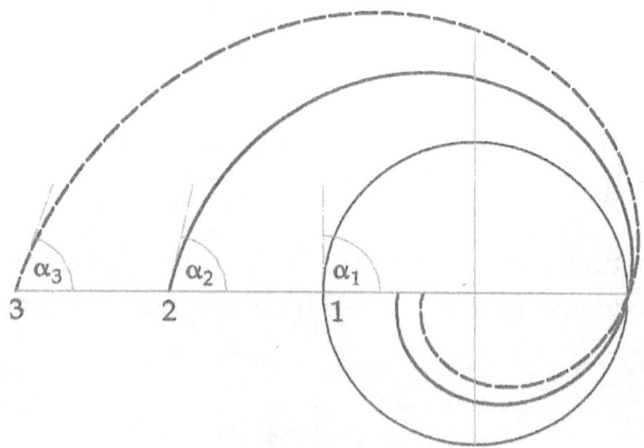

Referring to the drawing above, three values of K (1, 2 and 3) are indicated to provide an example for determining three corresponding values of α. The value of K determines the value of b

which in turn establishes the value of angle α in the standard polar equation $r = ae^{b\theta}$.

$$r = ae^{b\theta}$$

Substituting $a = 1$:

$$r = e^{b\theta}$$

Substituting $r = K$ when $\theta = \pi$,

$$K = e^{b\pi}$$

$$\ln K = b\pi$$

$$\frac{\ln K}{\pi} = b$$

$$\tan \alpha = \frac{1}{b} \quad \text{(See appendix L)}$$

$$\tan \alpha = \frac{1}{\dfrac{\ln K}{\pi}}$$

$$\tan \alpha = \frac{\pi}{\ln K}$$

If, for example, $K = 3$:

$$\tan \alpha_3 = \frac{\pi}{\ln 3}$$

$$\tan \alpha_3 = 2.859$$

$$\alpha_3 = \tan^{-1} 2.859$$

$$\alpha_3 = 70.725°$$

Angles α_2 (77.557°) and α_1 (90°) are determined in a similar manner. The pitch of the spiral at $K = 3$ is equal to 90° − 70.725 = 19.972°. The pitch of the spiral at $K = 2$ is 90°−77.557° = 12.442°. The pitch of the spiral at K = 1 is 90° − 90° = 0 and the spiral becomes a circle.

The curve of a logarithmic spiral can be approximated with a rotation of geometrically constructed quadrants of circles as related to eight equally spaced radius vectors.

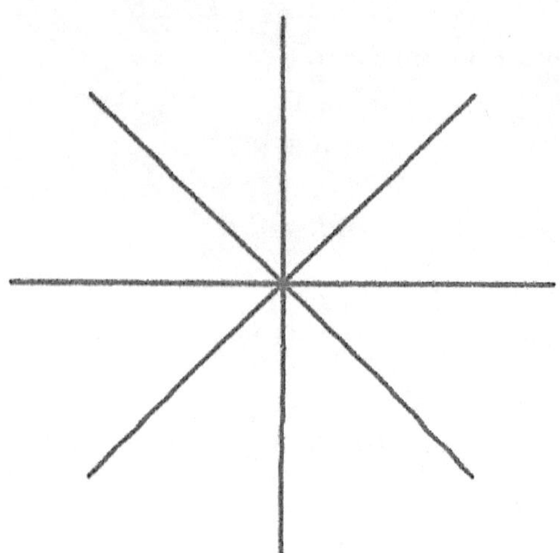

Eight equal radius vectors are drawn equally spaced.

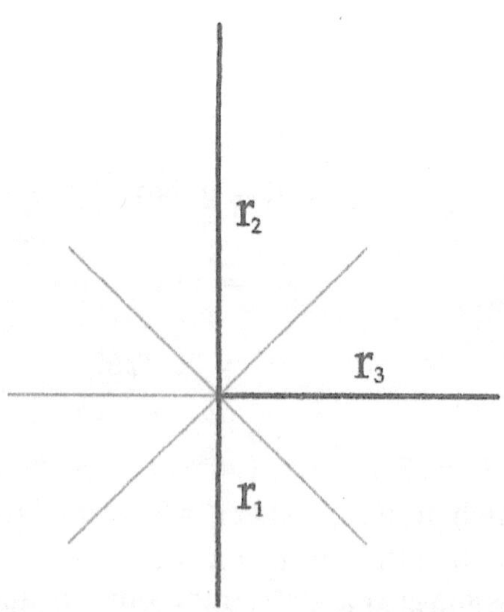

Radius vectors r_1, r_2 and r_3 are established as shown ($r_3 = \sqrt{r_1 \times r_2}$).

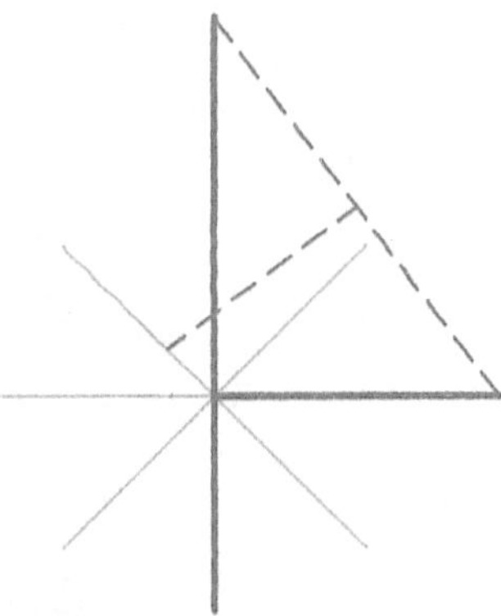

A dashed line is drawn from the ends of radius vectors r_3 and r_2. Another dashed line is drawn perpendicular to the midpoint of that line, and extended to the oblique radius vector to locate the center of an arc.

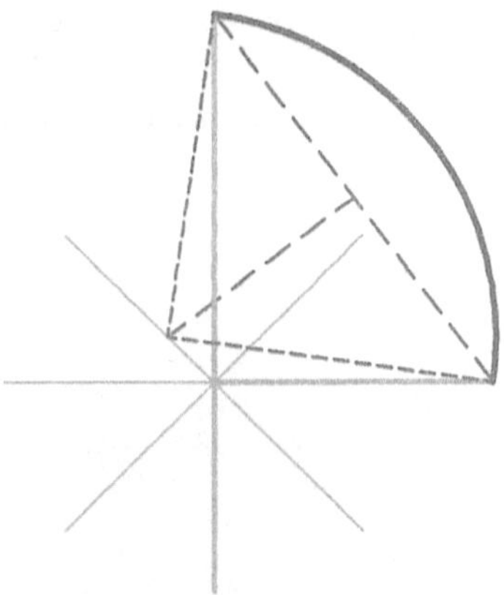

An arc is drawn from the ends of radius vectors r_3 and r_2.

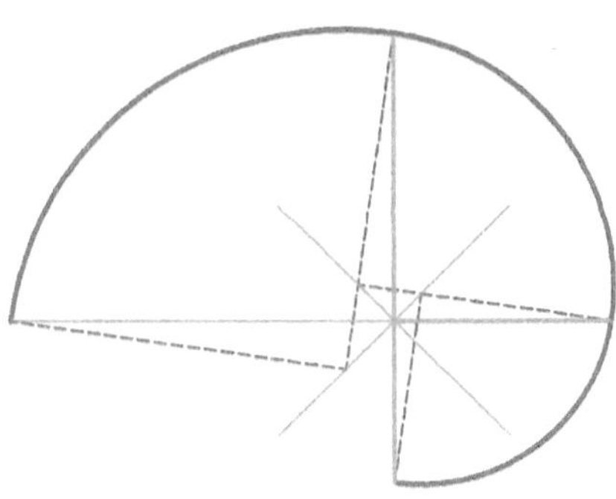

Two more arcs (one smaller, the other larger) are added in a similar manner.

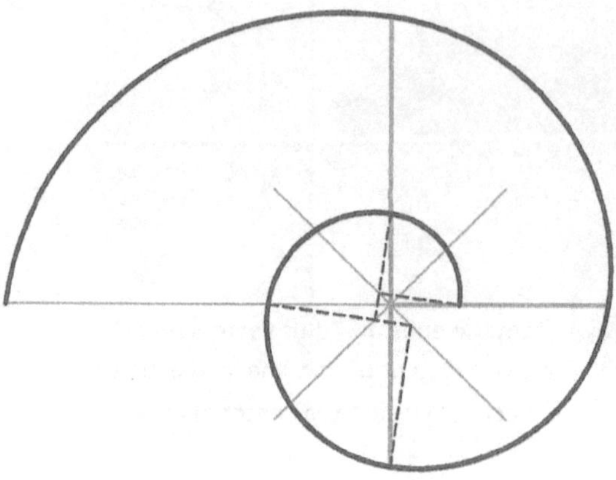

Three smaller arcs are added in a similar manner.

Logarithmic Spirals as Related to Rectangles

Since the shape of a logarithmic spiral can be established by a ratio, and since the proportions of any rectangle are also established by a ratio, (the comparison of the long side to the short side or vise versa), a relationship exists between a logarithmic spiral and a rectangle.

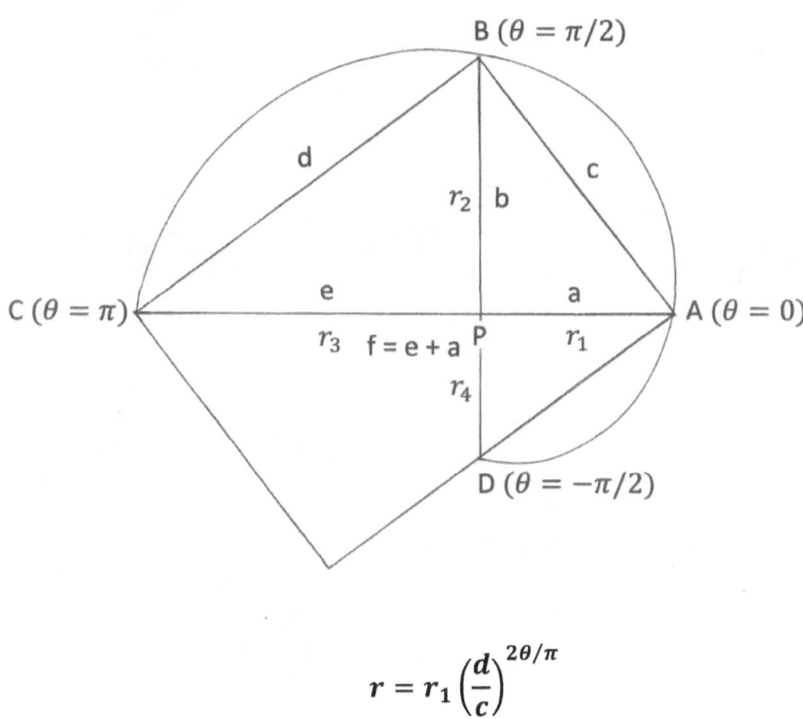

$$r = r_1 \left(\frac{d}{c}\right)^{2\theta/\pi}$$

In the diagram above, the logarithmic spiral coincides with the corners of a rectangle whose dimensions are c and d. Triangle *APB* is proportional to triangles *ABC* and *BPC*. Additionally, angles *APB*, *BPC* and *ABC* are right angles (90°). Therefore, the shape of a logarithmic spiral is related to the right triangle (which is one half of a rectangle) whose apexes touch the curve, and whose hypotenuse passes through the pole *P*. The polar equation for the spiral (see above) in terms of the ratio of the sides c and d of its accompanying rectangle is derived as follows. Note that this equation is different from the equation derived on page 58: radius vector a is not equal to one:

$$PA = r_1$$

$$PC = r_2$$

$$PB = r_3$$

$$\frac{r_1}{r_3} = \frac{c}{d}$$

$$\left(\frac{d}{c}\right) r_1 = r_3$$

$$\frac{r_2}{r_3} = \frac{d}{c}$$

$$\left(\frac{c}{d}\right) r_2 = r_3$$

$$\left(\frac{d}{c}\right) r_1 = \left(\frac{c}{d}\right) r_2$$

$$\frac{d^2}{c^2} = \frac{r_2}{r_1}$$

$$r_2 = r_1 \left(\frac{d^2}{c^2}\right)$$

Substituting $r_1 \left(\frac{d^2}{c^2}\right)$ for r_2 in the modified equation $r = r_2{}^{\theta/\pi}$ (see page 58):

$$r = r_1 \left(\frac{d^2}{c^2}\right)^{\theta/\pi}$$

$$r = r_1 \left(\frac{d}{c}\right)^{2\theta/\pi}$$

If c and d are given specific values, it can be shown that $a = c^2/\sqrt{c^2 + d^2}$ (see appendix V). Hence the equation for the spiral in terms of the dimensions c and d of the rectangle becomes:

$$r = \left(\frac{c^2}{\sqrt{c^2 + d^2}}\right)\left(\frac{d}{c}\right)^{2\theta/\pi}$$

The following equations represent dimensions of a 3/4 ratio spiral obtained with polar equations or proportional ratios:

$$\mathbf{a} = \frac{c^2}{\sqrt{d^2 + c^2}} = \frac{3^2}{\sqrt{4^2 + 3^2}} = \frac{9}{5} = \mathbf{1.8}$$

$$a = r_1 \left(\frac{d}{c}\right)^{2\theta/\pi}$$

$$= r_1 \left(\frac{4}{3}\right)^{2\times 0/\pi}$$

$$\mathbf{a} = 1.8 \left(\frac{4}{3}\right)^{0} = \mathbf{1.8}$$

. . . .

$$\frac{b}{d} = \frac{a}{c}$$

$$\frac{b}{4} = \frac{a}{3}$$

$$b = \frac{4}{3}a$$

$$\mathbf{b} = \frac{4}{3}\times 1.8 = \mathbf{2.4}$$

. . . .

$$b^2 + a^2 = c^2$$

$$b = \sqrt{c^2 - a^2}$$

$$= \sqrt{3^2 - 1.8^2}$$

$$b = \sqrt{5.76} = 2.4$$

. . . .

$$b = \sqrt{e \times a}$$

$$= \sqrt{3.2 \times 1.8}$$

$$b = \sqrt{5.76} = 2.4$$

. . . .

$$b = r_2 = 1.8 \left(\frac{4}{3}\right)^{2\theta/\pi}$$

$$= 1.8 \left(\frac{4}{3}\right)^{2\pi/2/\pi}$$

$$\boldsymbol{b} = 1.8 \left(\frac{4}{3}\right)^{1} = 2.4$$

. . . .

$$e = f - a$$

$$e = 5 - 1.8 = 3.2$$

. . . .

$$e = r_3 = r_1 \left(\frac{4}{3}\right)^{2\pi/\pi}$$

$$e = 1.8 \left(\frac{4}{3}\right)^{2} = 3.2$$

. . . .

$$r_4 = 1.8 \left(\frac{4}{3}\right)^{-2\theta/\pi}$$

$$= 1.8 \left(\frac{4}{3}\right)^{-2\pi/2/\pi}$$

$$= 1.8 \left(\frac{4}{3}\right)^{-1}$$

$$r_4 = 1.8 \times \left(\frac{3}{4}\right) = 1.35$$

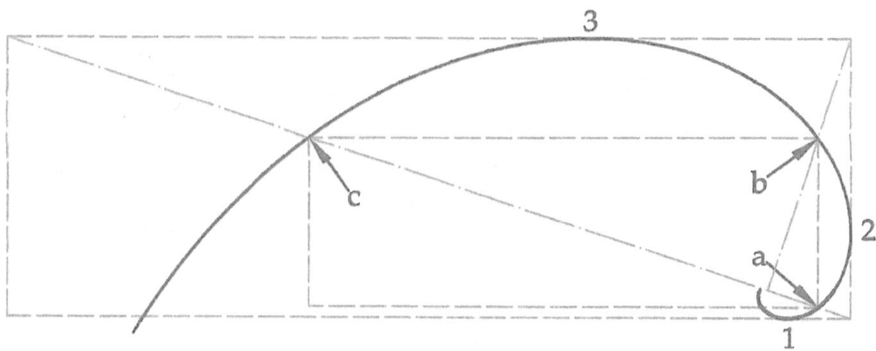

The spiral will always touch three corners of the inscribed rectangle, and will be tangent at either three or four points of the circumscribed rectangle. For example, with regard to the spiral shown above, the spiral touches three sides (1, 2 and 3) of the circumscribed rectangle, and three corners (*a*, *b* and *c*) of the inscribed rectangle.

The spiral drawn below is tangent to four sides of the circumscribed rectangle, and touches three corners of the inscribed rectangle. Note that the diagonal of the rectangle associated with a spiral always passes through the pole point of the spiral:

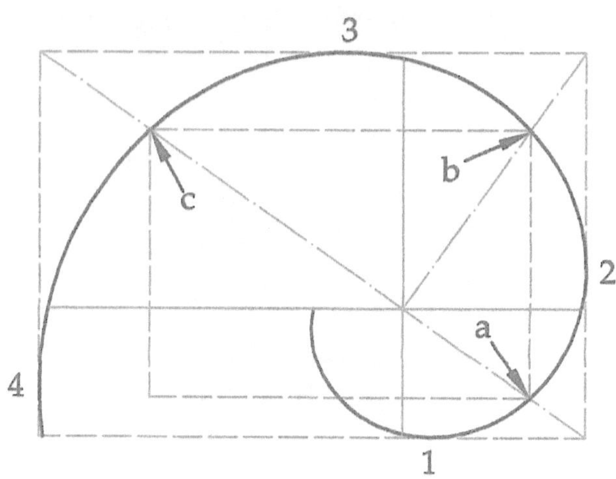

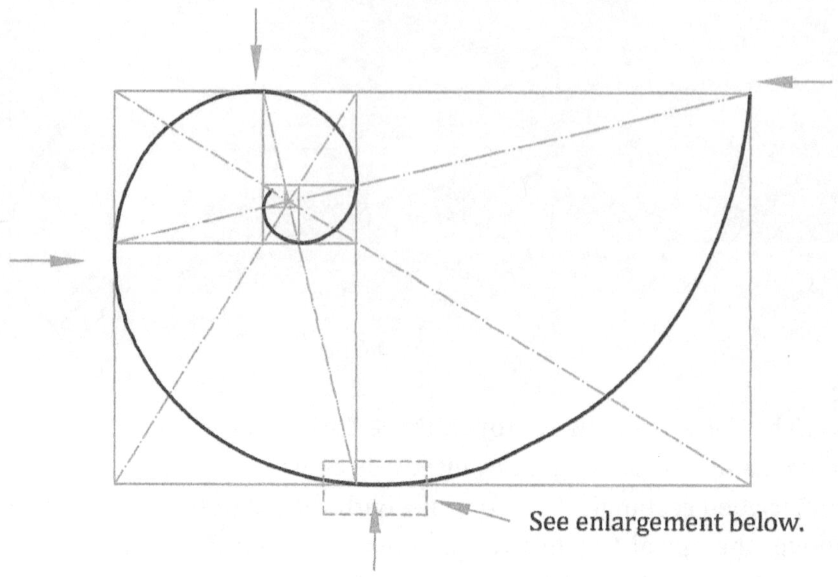

The spiral generated from similar points of golden rectangles as shown above, appears to be tangent to the sides of the rectangles at the corners of the rotating squares. However, this is not the case. If the spiral is coincidental with the corner of each respective square, the curve of the spiral will extend slightly outside the line of the rectangle, beyond the corners as indicated by the arrows, and as shown in the enlargement below. (See following drawings and Appendix *O*).

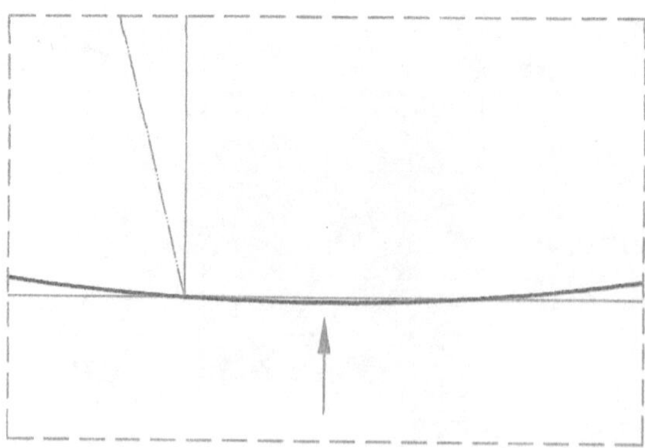

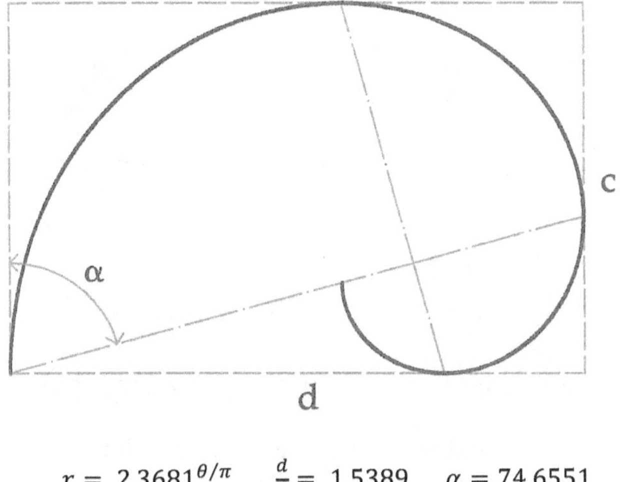

$$r = 2.3681^{\theta/\pi} \qquad \frac{d}{c} = 1.5389 \qquad \alpha = 74.6551$$

The spiral that is tangent at the corner of a rectangle is generated by the equation $r = 2.3681^{\theta/\pi}$. The ratio $d{:}c$ is equal to 1.5389. The tangential angle α is 74.6551°. (See appendix O).

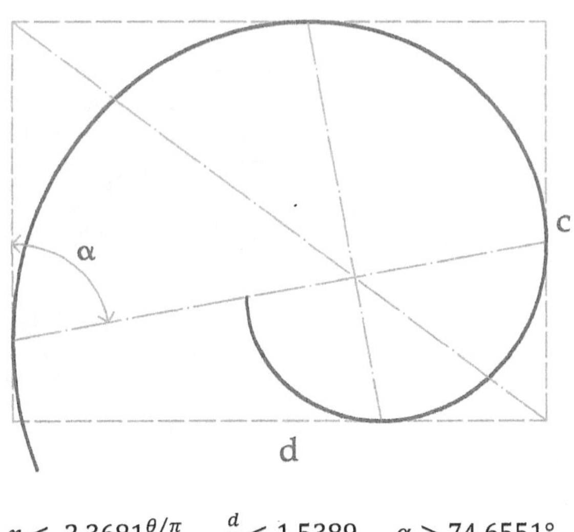

$$r < 2.3681^{\theta/\pi} \qquad \frac{d}{c} < 1.5389 \qquad \alpha > 74.6551°$$

If the ratio $d{:}c$ is less than 1.5389 α is greater than 74.6551°, and four sides will be tangent to the spiral:

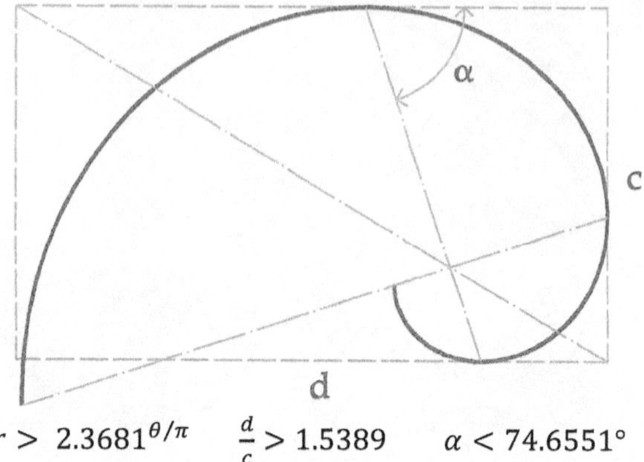

$$r > 2.3681^{\theta/\pi} \qquad \frac{d}{c} > 1.5389 \qquad \alpha < 74.6551°$$

If the ratio *d:c* is greater than 1.5389 α is less than 74.6551°, and three sides will be tangent to the spiral.

The Fibonacci spiral (shown below) is generated by a rotation of squares representing Fibonacci numbers beginning with 1. The curve is produced by connecting similar corners (solid line) or by tangentially touching similar sides of the squares (dashed line):

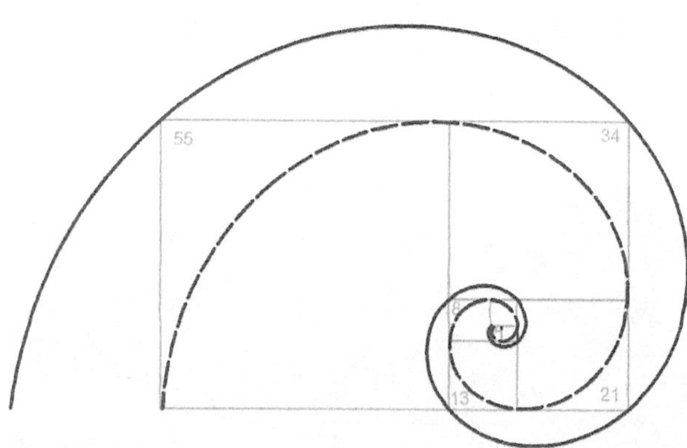

The equation defining the spiral varies according to the adjacent terms of the rotating sequence of Fibonacci numbers. That is to say, the ratio of the sides of the rectangle defining the curve at each square is slightly different: 3/5, 13/8, 21/34, and so forth. For example, the equation of the curve at its beginning is:

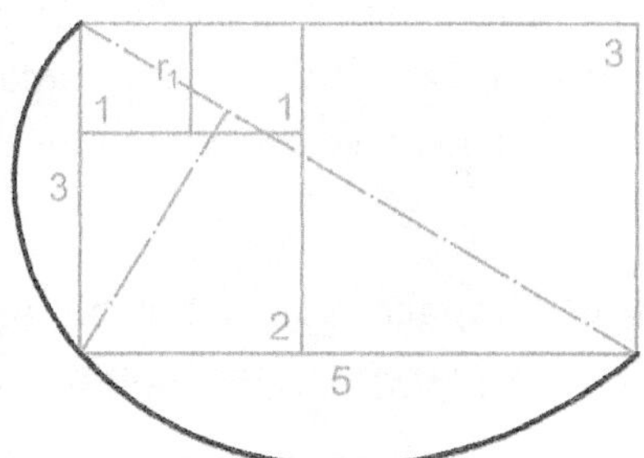

$$r = r_1 \left(\frac{5}{3}\right)^{2\theta/\pi} = \left(\frac{3^2}{\sqrt{3^2 + 5^2}}\right)\left(\frac{5}{3}\right)^{2\theta/\pi} = 1.5435(1.6667)^{2\theta/\pi}$$

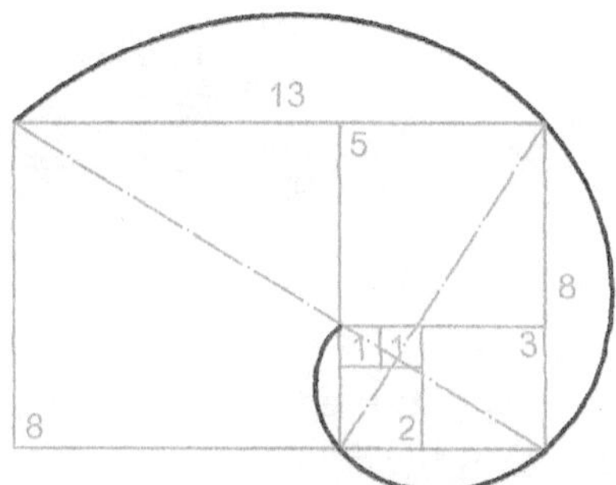

$$r = r_1 \left(\frac{13}{8}\right)^{2\theta/\pi} = \left(\frac{8^2}{\sqrt{8^2 + 13^2}}\right)\left(\frac{13}{8}\right)^{2\theta/\pi} = 4.1928(1.625)^{2\theta/\pi}$$

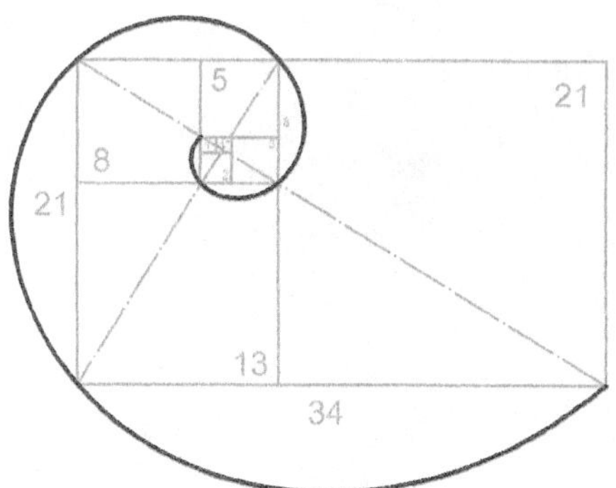

$$r = r_1 \left(\frac{34}{21}\right)^{2\theta/\pi} = \left(\frac{21^2}{\sqrt{21^2 + 34^2}}\right)\left(\frac{34}{21}\right)^{2\theta/\pi} = 11.0354(1.6190)^{2\theta/\pi}$$

It is apparent that the expansion of the spiral can continue indefinitely as the ratio d:c (the ratio of the sides of the rectangle made up of the squares) converges on the value of ϕ.

The curve of the Fibonacci spiral can be approximated using the arcs of circles (each square is a quadrant of a circle):

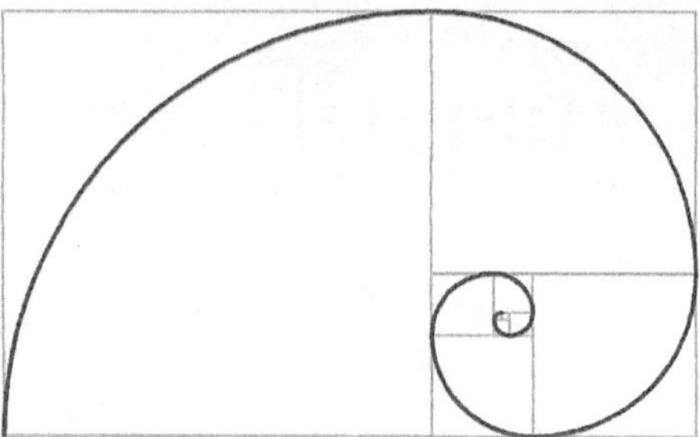

The curved portion of the edge of the cockle shell illustrated below is a portion of a logarithmic spiral proportioned according to the 4:3 ratio. Additionally, the entire shell fits into a rectangle proportioned according to a 9:10 ratio (from the sequence, 1/2, 2/3, 3/4, 4/5 ... 8/9, 9/10, 10/11 ...):

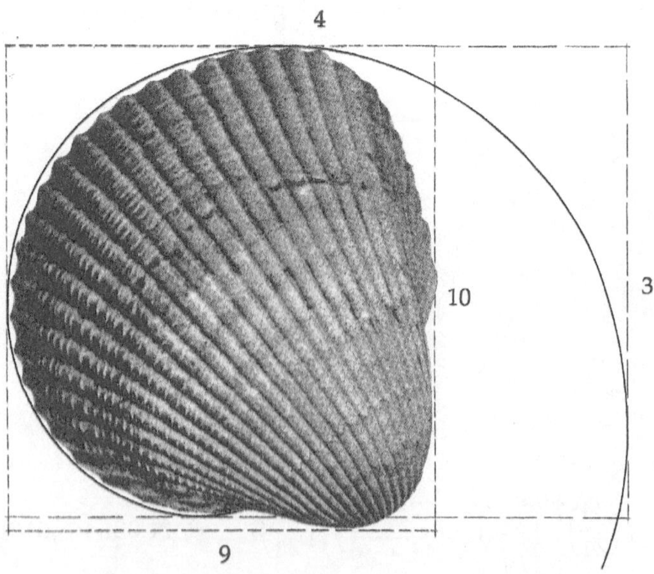

The curved portion of the longitudinal profile of each half of the cockle shell is also a logarithmic spiral. The curve is generated from a rectangle proportioned according to a 3:1 ratio:

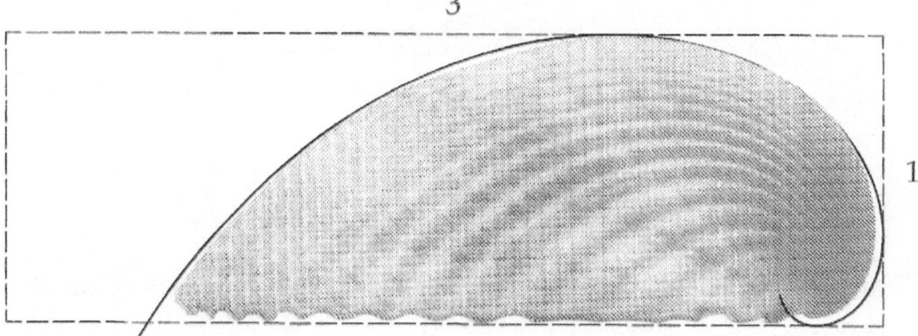

If the dimensions are 3 units and 1 unit as shown, the formula for the spiral (as previously derived) is:

$$r = \frac{d^{2\theta/\pi}}{\sqrt{1 + d^2}}$$

$$r = (0.3162) \times 9^{\theta/\pi}$$

The shell of a common garden snail (below) can be inscribed in a rectangle whose sides are proportioned according to either a 10:9 or 11:10 ratio (the ratios differ by only 1/100: 11/10 = 1.1; 10/9 = 1.11). Note that the spiral of the snail shell increases in a clockwise manner (as would be the case with the other half of the cockle shell).

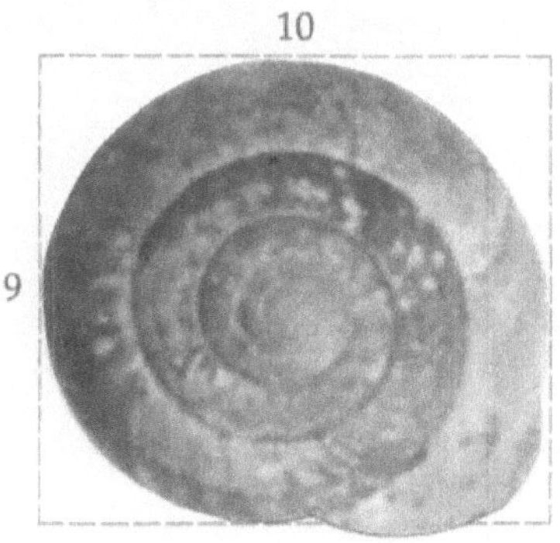

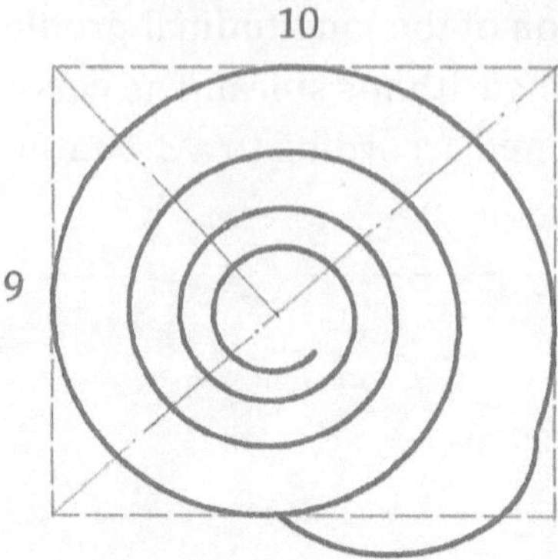

The formula for the spiral (as previously derived) based on the rectangle with dimensions 10 and 9 is,

$$r = \left(\frac{c^2}{\sqrt{c^2 + d^2}}\right)\left(\frac{d}{c}\right)^{2\theta/\pi}$$

resulting in:

$$r = (6.0207)\times 1.23^{\theta/\pi}$$

Note that the outer coil of the shell matches the algebraic, logarithmic spiral exactly. However, the inner coils deviate from the mathematical curve considerably as shown in the diagram below, where the actual curve of the shell is indicated by the dashed line.

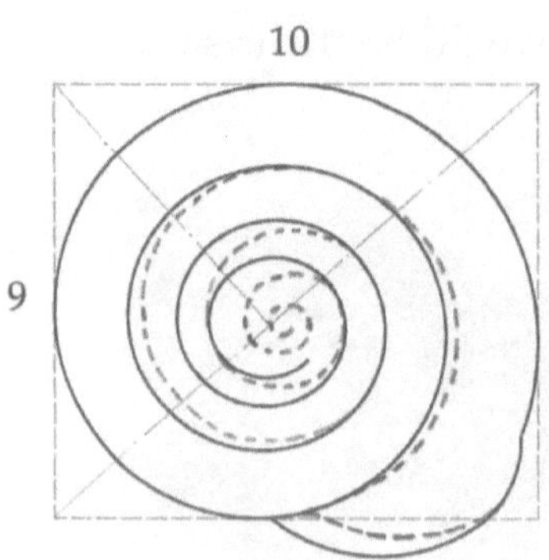

The Logarithmic Spiral of the Shell of the Chambered Nautilus

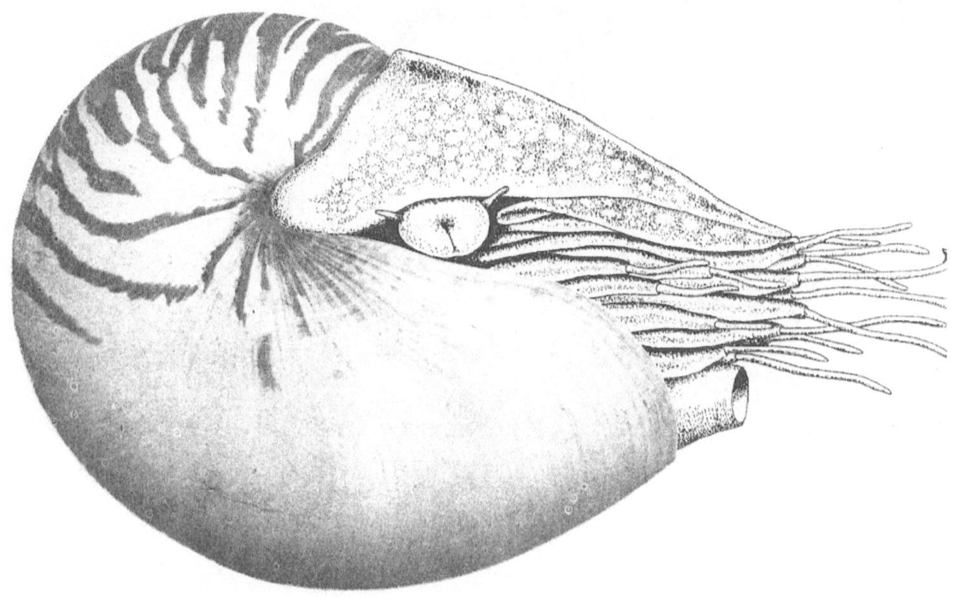

Nautilus pompilius

The composite photograph/sketch above shows a chambered nautilus in its normal position of neutral buoyancy as it would appear in its natural habitat of between 700 and 900 feet below the surface of the sea. It can safely descend to a depth of 1500 feet below the surface.[*]

The following drawings and photographs illustrate a two-dimensional analysis of the curve of the cross-section of the shell.

---

[*] Peter Douglas Ward, *In Search of Nautilus,* (Simon Snd Schuster, 1988, pages 14,100, 129 and 155.

The photograph above, of a nautilus shell that has been sawn in half reveals its exquisite internal structure: a series of pearly, calcareous septa, (forming three-dimensionally curved chambers) arranged according to the mechanics and principles of a logarithmic spiral. In a living animal, the chambers contain a saline fluid whose density is regulated by osmosis enabling the creature to maintain the buoyancy needed to negotiate the ocean's depths.

The same N. pompilius shell as related to a rectangle proportioned according to a 4:3 ratio:

$$\frac{d}{c} = \frac{4}{3}$$

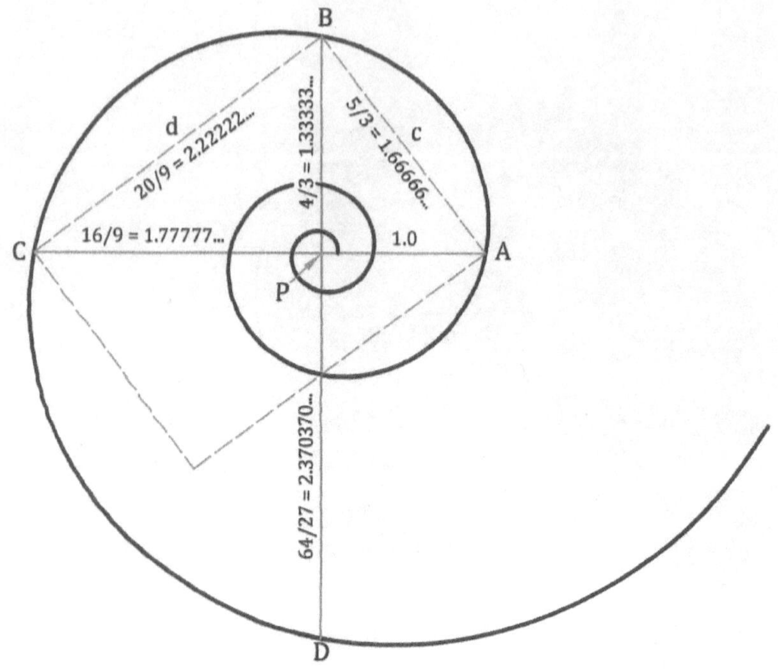

The logarithmic spiral above is generated from a rectangle whose proportions are 4:3. **If *PA* = 1**, the modified polar equation for the spiral is:

$$r = \left(\frac{4}{3}\right)^{2\theta/\pi}$$

In terms of the standard polar equation the formula is:

$$r = e^{0.18314\,\theta}$$

These equations result in the fascinating repeating decimal values shown for *PB, PC, PD, AB* and *BC*:

1.33333..., 1.77777..., 2.370370..., 1.66666..., and 1.22222...

The same mathematically constructed logarithmic spiral curve superimposed over the cross section of a nautilus shell.

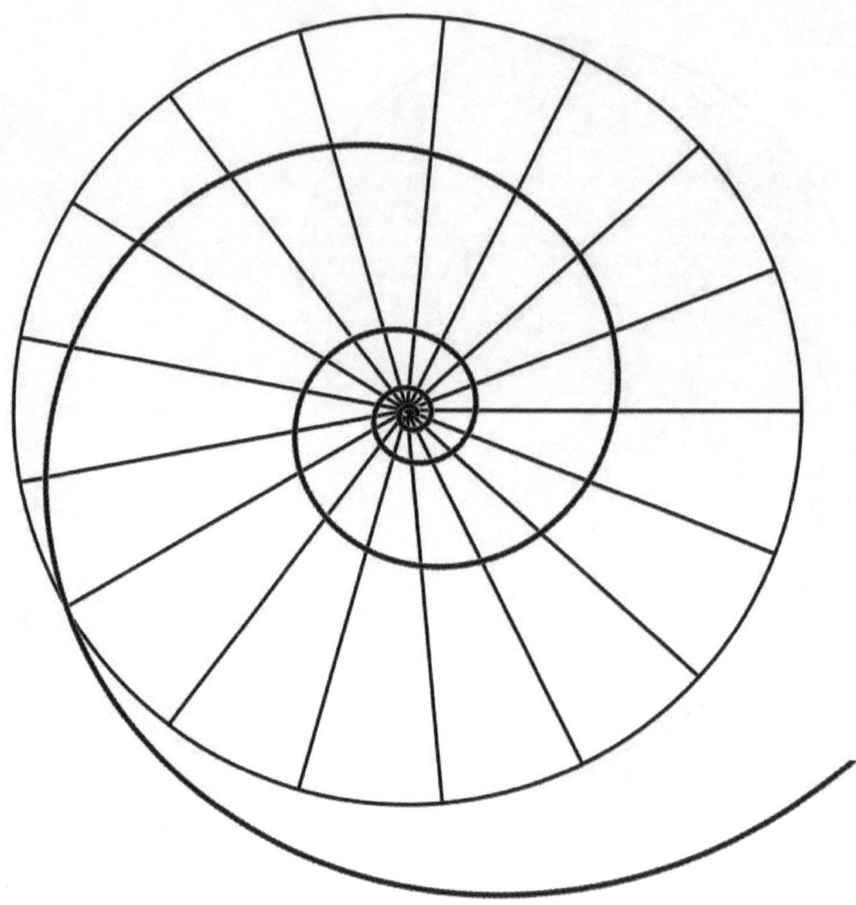

This drawing shows seventeen equally spaced radius vectors as related to the logarithmic spiral curve.

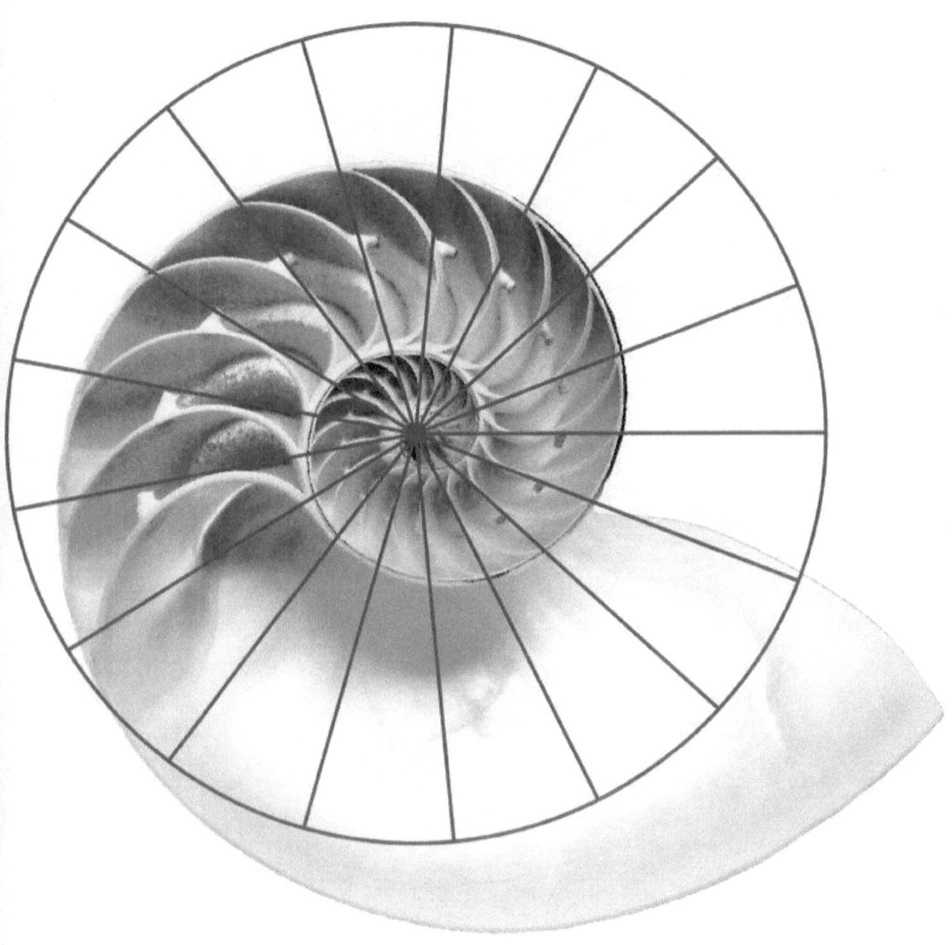

The seventeen equally spaced radius vectors are shown superimposed over the cross section of the nautilus. Remarkably, the majority of the walls of the chambers (beginning with the outermost chamber) align tangentially to the radius vectors.

In this drawing, the curves of the chamber walls are positioned tangentially to each of the seventeen radius vectors using a portion of the curve of the logarithmic spiral at a reduced scale.

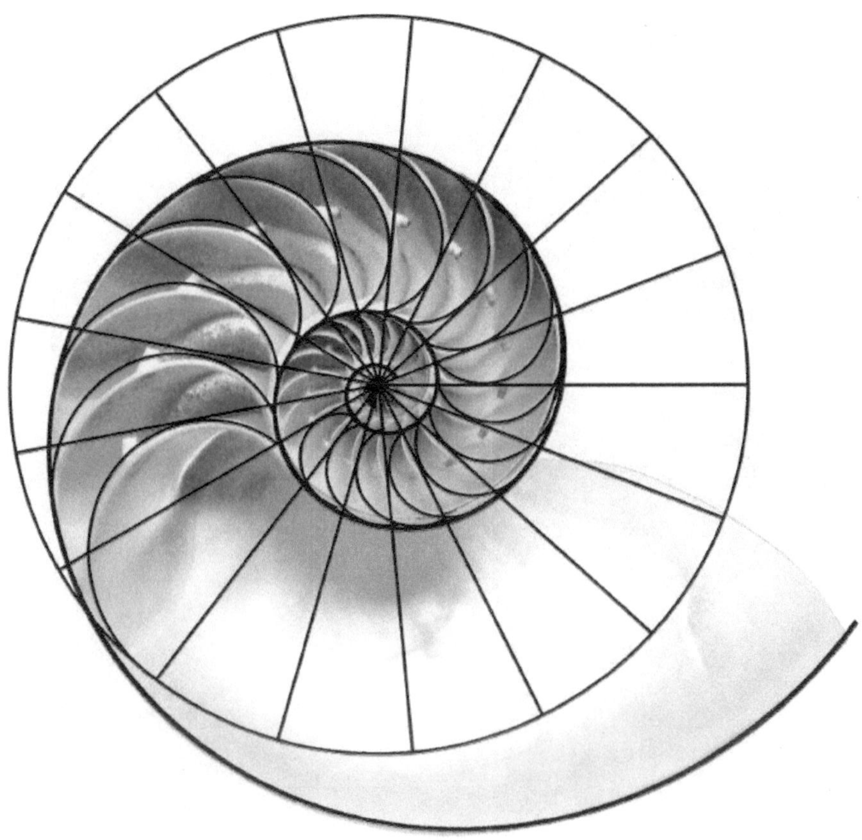

The mathematically constructed spiral curve, and the portion of the spiral curve that creates the curve of the chamber walls (see preceding page) are shown superimposed over the actual spiral curve and chambers of the nautilus. Many of the chamber walls coincide remarkably with the geometrically constructed curve of the chambers.

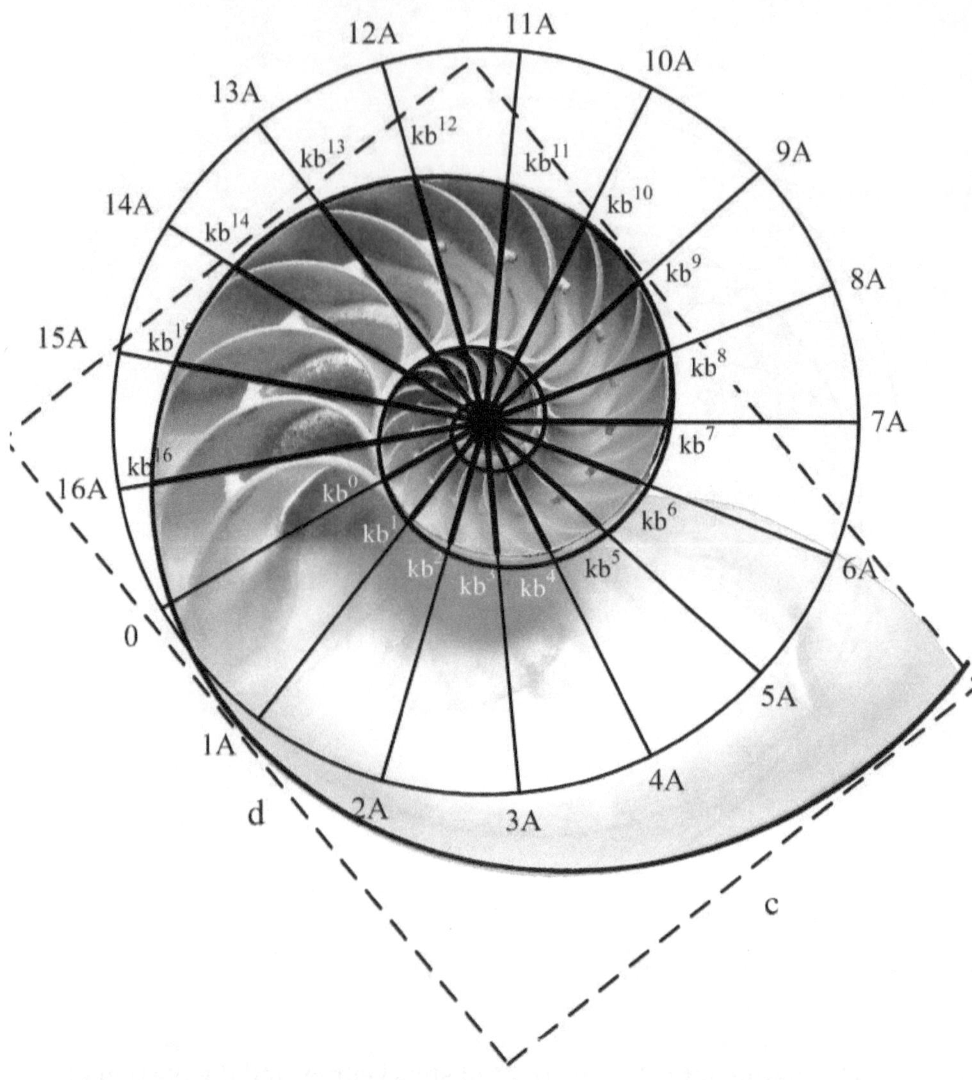

This composite drawing illustrates the salient features of the
relationship between the nautilus (as related to a two-dimensional
photograph) and its underlying mathematics. The seventeen segments of a circle representing the arithmetic progression of the radius vectors are labeled *1A, 2A, 3A* and so forth. The magnitudes of

the radius vectors, in geometric progression relating to the walls of the chambers, are labeled kb^0, kb^1, kb^2, kb^3, and so forth.

The letter A designates the constant angle between the radius vectors representing the arithmetic progression. Thus the value of A in radians is equal to one seventeenth the circumference of the circle: $2\pi/17$. If $k = 1$, the radius vectors expand according to the exponentiation of b. The radius vector equal to $b^0 = 1$ is located at zero radians. The value of the radius vector $b^1 = b$, located at $2\pi/17$ radians, is determined using the formula,

$$r = \left(\frac{4}{3}\right)^{2\theta/\pi}$$

$$= \left(\frac{16}{9}\right)^{\theta/\pi}$$

Substituting $\theta = 2\pi/17$ $\qquad b = (1.7777)^{2\pi/17/\pi}$

$$= (1.7777)^{2/17}$$

$$= 1.0700$$

Hence the chambers expand in two dimensions (area) according to the exponentiation of b:

$$b^0, \quad b^1, \quad b^2, \quad b^3, \quad b^4, \quad b^5 \dots$$

$$1.0, \quad 1.0700, \quad 1.1449, \quad 1.2250, \quad 1.3108, \quad 1.4026,$$

It can be assumed that the chambers expand in three dimensions (volume) also according to a geometric progression, possibly the same progression that expands the area.

Epilogue

This two-dimensional analysis of the shell of the chambered nautilus introduces four questions that deserve comment. The first is, *"How is it known that the curve of every nautilus shell is based on a 4:3 ratio, or that the chambers are aligned according to sixteen or seventeen radius vectors?"* The obvious answer is that it is not known. However, if every observed shell reveals the same relationships, it can reasonably be assumed that all shells will be similar.

The second question concerns the relationship between the mathematics underlying the geometry of the observed shells and the actual expression of the mathematics as a form: *"Why are the shells not an exact expression of pure geometry?"* For example, the septa of the smallest chambers of the shell presented do not fall exactly on the seventeen radius vectors. Neither is the profile of the shell's spiral a perfect logarithmic curve (albeit nearly perfect). In a philosophical sense, imperfection (as related to pure mathematics) is an important aspect of all living things. Because of "imperfection" none of the infinite variety of biological entities is exactly alike. Each expresses its own identity based on "imperfection." Yet within this aspect of individuality, there exists an underlying unity of mathematics.

Not only is this evident in the nautilus shell, but also in the black-eyed Susan as previously described. Even though the underlying basis for the expression of the flower appears to be specific Fibonacci relationships, all flowers are slightly different. The obvious reason for these differences is the infinite number of variables encountered during the growth process. Such factors as nutrition, temperature, sunlight, moisture, motion and gravity (including the gravity forces of planetary influences) affect growth, substance and subsequent expression of a life form. More evidence of these variables can be seen in the comparison of the shells in the following illustration.[†]

The mollusk shells on the left are real. Those on the
right are virtual, generated by a computer.

The photographs on the left are of real shells expressing identity and character based on an infinite resource of potentiality that provides an unlimited variety of expression. Those on the right are from a computer monitor having been generated by a computer using precise, abstract, programming instructions. However, the computer images appear lifeless with no nuances of "personality." They are the result of limited mathematical formulas and the restriction of computing capacity. Indeed, life as we know it would be inconceivable if every biological entity was a perfect expression of

---

[†] Carl Zimmer, *Shell Game, Discover* Magazine, May 1992 p. 37-43. Photographs of the real shells are by Guiseppe Mazza. The computer images of the shells were created by a team led by Przemyslaw Prusinkiewicz of the University of Calgary.

precise mathematics, identical to every other, lacking the broad range of infinite capability of expression.

The third question is more profound. It relates to not only the nautilus shell but also to every example of geometry discussed in this book regarding either a natural form or an abstract mathematical expression. It is apparent that nature does not exist in a chaotic, random manner. Rather, nature exists as expressions of coherence, orderliness and fundamental truths as opposed to random coincidences. Consider for example, the different methods determining the value (to nine decimal places) of the golden proportion:

(1) the solution to the quadratic equation $\phi^2 = \phi + 1$;

$$\phi = \frac{1 + \sqrt{5}}{2} = 1.618033989$$

(2) the ratio of the 25th and 24th terms of the Fibonacci sequence;

$$\phi = \frac{75,025}{46,368} = 1.618033989$$

3) the ratio of the 24th and 25th terms of the Fibonacci sequence expressing the reciprocal of phi;

$$\phi - 1 = \frac{1}{\phi} = \frac{46,368}{75,025} = 0.618033989 = 1.618033989 - 1.0$$

(4) the ratio of the sines of the greater and lesser angles of the golden triangle;

$$\phi = \frac{\sin 72°}{\sin 36°} = 1.618033989$$

(5) the value of the 22th iteration of the infinitely continuing fraction;

$$\phi = 1 + \cfrac{1}{1 + \cfrac{1}{1 + \cfrac{1}{1 + \cdots}}} = 1.618033989$$

(6) the ratio of the long side to the short side of a golden rectangle:

$$\phi = \frac{\frac{1 + \sqrt{5}}{2}}{1}$$

$$\phi = 1.618033989$$

These six different methods determining the value of phi can be considered a coincidence, but more likely, the numerical value is the result of astounding, if not extraordinary, universal intelligence. The fact that $\phi^2 = \phi + 1$ (resulting in $1/\phi = \phi - 1$) is even more extraordinary—the "icing on the cake" so to speak.

Most dictionaries define intelligence in terms of an individual, such as an individual's self-awareness, ability to learn, understand, reason or communicate. However, the reality of individual intelligence can also be considered to be the reality of universal intelligence: the ability of any individual to understand and comprehend nature's orderliness, coherence and infinitely diverse relationships—nature's intelligence.

Having accepted the existence of the reality of nature's intelligence—a very small portion of which is exhibited by the limited examples in this book—it is not implausible to conclude that the entire universe is an expression of intelligence, and that this intelligence is creative. This prompts the logical third question which has been posed throughout the ages: *"Where does creative intelligence come from?"* What is its source? Enlightened individuals suggest that the source is an unmanifest field of pure consciousness, existing everywhere, omnipresent with no beginning or end.

If this field exists, it must be able to be experienced. Fortunately, this is indeed the case. The human brain is so structured that it can subjectively experience pure consciousness: described by those who experience it as a lively field of pure awareness, devoid of thought—infinite, with no boundaries.

I came to know this field of lively awareness—as have millions of others worldwide—by means of the Transcendental Meditation® (TM®) technique as taught by the late Maharishi Mahesh Yogi. His simple, effortless technique enables anyone (who can think a thought) to transcend meditation (go beyond the thinking process). The ability to transcend thought enables the mind to experience the unbounded field of pure consciousness.

The educational program established by Maharishi to learn the technique includes a complete understanding of the experience of transcending and its relation to human physiology, broadened awareness, and enlightenment. Having taken the TM® course, and having been a practitioner of this technique for over forty years, it is my great joy to have had the opportunity to research, understand and produce the content of this book—examples of nature's ratios, progressions and spirals as related to plants and shells—in light of broadened awareness.

The fourth question, "*How do the expressions of nature's infinite shapes, forms and functioning get formulated?*" is the result of having addressed the third question. For example, how does the DNA, which controls the development and functioning of living organisms, come to be arranged from a field of consciousness devoid of boundaries; how does the nautilus know how or when to alter the density of the saline fluid in its chambers in order to ascend or descend in the depths of the ocean? How do mathematical formulations come about to relate to natural organisms or the functioning of matter?

Many think the living organisms of the Creation are the result of random coincidence enabling minute changes to occur over eons of time. Mathematical equations are thought to be attributable to those who derive them as opposed to originating in a state of potentiality, waiting to be unfolded at the time they are needed.

However, many others believe that the formulation and expression of any aspect of nature's intelligence, including the timelessness of mathematics, is nothing more than Divine.

Contents of the Appendix

Appendix A:
Baseball Team Standings—Unequal Games Played

If, in the same league, the number of games played by any baseball team is different from the number of games played by the first place team, the standing of the teams (as related to the first place team) is calculated using a ratio. The number of games any team is behind the first place team is determined by taking the difference between the games won and lost of each of the two teams, adding these differences together, and dividing the sum by two.

For example, if the first place team has won 20 games and has lost 12 games (for a total of 32 games played), and the next best team (in second place) has won 18 games and lost 13 games (for a total of 31 games played), the difference between games won (of the two teams) is 2, and the difference between games lost is 1. Hence $(2 + 1)/2 = 3/2 = 1\frac{1}{2}$. Therefore, the next best team is $1\frac{1}{2}$ games out of first place.

If the last place team has won 9 games and lost 24 games (for a total of 33 games played), the last place team is $11\frac{1}{2}$ games out of first place: $(11+12)/2 = 23/2 = 11\frac{1}{2}$. That is to say, the last place team needs to add $11\frac{1}{2}$ winning games to its list of games won to approximate the ratio of games won to games played of the first place team. It should be noted of course, that teams do not play half games. The half games are only used to approximate the standing at any given time during the season with regard to unequal games played. At the end of the season, all teams will have played an equal number of games, negating the half games in the standings.

Appendix B: Properties of Proportional Ratios

Given the proportion $a/b = c/d$, (where a, b, c and d represent any positive numbers) it is easily shown that the product of the means is equal to the product of the extremes ($a{\times}d = b{\times}c$):

$$\frac{a}{b} = \frac{c}{d}$$

$$bd{\times}\frac{a}{b} = \frac{c}{d}{\times}bd$$

$$d{\times}a = c{\times}b$$

It can also be shown that:

$$\frac{b}{a} = \frac{d}{c}$$

$$\frac{a}{c} = \frac{b}{d}$$

$$\frac{a-b}{b} = \frac{c-d}{d}$$

$$\frac{a+b}{b} = \frac{c+d}{d}$$

$$\frac{a+b}{a-b} = \frac{c+d}{c-d}$$

Additionally,

$$\frac{\sqrt{a}}{a} = \frac{1}{\sqrt{a}}$$

$$\frac{a/b}{a} = \frac{1}{b}$$

$$\frac{a/b}{b} = \frac{a}{b^2}$$

$$\frac{a/b}{c} = \frac{a}{bc}$$

$$\frac{a}{a/b} = b$$

$$\frac{a}{b/a} = \frac{a^2}{b}$$

$$\frac{a}{b/c} = \frac{ac}{b}$$

The equation, $\frac{a-b}{b} = \frac{c-d}{d}$ is derived knowing that if a/b = c/d, then $ad = bc$.

$$ad - bd = bc - bd$$

$$d(a - b) = b(c - d)$$

$$\frac{d(a - b)}{b} = \frac{b(c - d)}{b}$$

$$\frac{d(a - b)}{b} = (c - d)$$

$$\frac{d(a - b)}{bd} = \frac{(c - d)}{d}$$

$$\frac{a - b}{b} = \frac{c - d}{d}$$

The derivation of $\frac{a+b}{b} = \frac{c+d}{d}$ is similar to the derivation of $\frac{a-b}{b} = \frac{c-d}{d}$ except that bd is <u>added</u> to each side of the equation $ad = bc$.

The equation $\frac{a+b}{a-b} = \frac{c+d}{c-d}$ is derived knowing $\frac{a-b}{b} = \frac{c-d}{d}$, and $\frac{a+b}{b} = \frac{c+d}{d}$:

$$\frac{a - b}{b} = \frac{c - d}{d}$$

$$d(a - b) = (c - d)b$$

$$d = \frac{(c - d)b}{a - b}$$

Since $\frac{a+b}{b} = \frac{c+d}{d}$,

$$\frac{a + b}{b} = \frac{c + d}{\frac{(c - d)b}{a - b}}$$

$$\frac{a + b}{b} = \frac{(c + d)(a - b)}{(c - d)b}$$

$$a + b = \frac{(c + d)(a - b)}{(c - d)}$$

$$(a + b)(c - d) = (c + d)(a - b)$$

$$\frac{(a + b)(c - d)}{(a - b)(c - d)} = \frac{(c + d)(a - b)}{(a - b)(c - d)}$$

$$\frac{(a + b)}{(a - b)} = \frac{(c + d)}{(c - d)}$$

Appendix C: **Properties of Exponents**

The following relationships involving exponents hold true providing a is a positive integer unequal to zero:

$$a^0 = 1$$

$$a^r \times a^s = a^{r+s}$$

$$\frac{a^r}{a^s} = a^{r-s}$$

$$(a^r)^s = a^{rs}$$

$$a^{-s} = \frac{1}{a^s}$$

$$\frac{\sqrt{a}}{a} = \frac{1}{\sqrt{a}}$$

$$\left(\frac{a}{b}\right)^r = \frac{a^r}{b^r}$$

$$a^{1/s} = \sqrt[s]{a}$$

$$a^{r/s} = \left(\sqrt[s]{a}\right)^r$$

If $n^s = a$: $\qquad\qquad r = \log_n a$

If $a^{1/s} = n$: $\qquad\qquad a = n^s$

If $a^{1/s} = n$: $\qquad\qquad \frac{1}{s} = \log_a n$

$$n^{\log_n s} = s$$

$$e^{\ln s} = s$$

Appendix D: The Quadratic Equation

The two solutions for x in the general form of the quadratic equation $ax^2 + bx + c = 0$, are as follows:

$$x_1 = \frac{-b + \sqrt{b^2 - 4ac}}{2a}$$

$$x_2 = \frac{-b - \sqrt{b^2 - 4ac}}{2a}$$

Hence in the quadratic equation, $a\phi^2 - b\phi - c = 0$, if a, b, and $c = 1$, such that $\phi^2 - \phi - 1 = 0$:

$$\phi_1 = \frac{1 + \sqrt{5}}{2} = 1.61803\ldots$$

$$\phi_2 = \frac{1 - \sqrt{5}}{2} = 0.61803\ldots$$

With regard to the two solutions (x_1 and x_2) of the quadratic equation:

$$x_1 + x_2 = -\frac{b}{a}$$

$$x_1 \times x_2 = \frac{c}{a}$$

And therefore: $\phi_1 + \phi_2 = 1$

$$\phi_1 \times \phi_2 = -1$$

The solution for x in the standard quadratic equation begins by knowing $(x + d)^2 = x^2 + 2xd + d^2$. Hence, with regard to the equation:

$$ax^2 + bx + c = 0$$

Dividing both sides of the equation by a:

$$x^2 + \frac{bx}{a} + \frac{c}{a} = 0$$

$$x^2 + \frac{bx}{a} = -\frac{c}{a}$$

$$x^2 + \frac{bx}{a} + \left(\frac{b}{2a}\right)^2 = -\frac{c}{a} + \left(\frac{b}{2a}\right)^2$$

$$\left(x + \frac{b}{2a}\right)^2 = -\frac{c}{a} + \frac{b^2}{4a^2}$$

$$= \frac{b^2}{4a^2} - \frac{c}{a}$$

$$= \frac{b^2}{4a^2} - \frac{c4a}{4a^2}$$

$$= \frac{b^2-4ac}{4a^2}$$

$$x + \frac{b}{2a} = \pm\sqrt{\frac{b^2-4ac}{4a^2}}$$

$$x + \frac{b}{2a} = \pm\frac{\sqrt{b^2-4ac}}{2a}$$

$$x = -\frac{b}{2a} \pm \frac{\sqrt{b^2-4ac}}{2a}$$

$$x = \frac{-b\pm\sqrt{b^2-4ac}}{2a}$$

Appendix E: Proof of the Division of a line into the phi proportion.

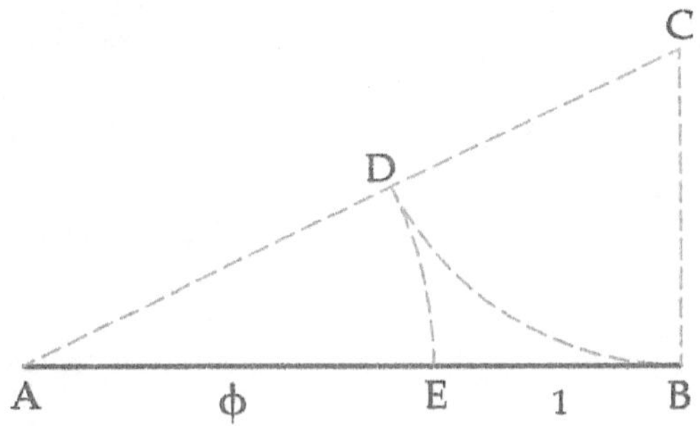

The ratio $AE:EB = \phi:1$ is derived using the Pythagorean theorem to prove that $(AE)^2 - AE - 1 = 0,$ the quadratic equation defining phi where BC is perpendicular to AB:

$$BC = \frac{AB}{2}$$

$$(AB)^2 + (BC)^2 = (AC)^2$$

$$(AE + 1)^2 + \left(\frac{AE+1}{2}\right)^2 = \left(AE + \frac{AE+1}{2}\right)^2$$

$$(AE)^2 + 2(AE) + 1 + \left(\frac{AE+1}{2}\right)^2 = (AE)^2 + 2(AE)\left(\frac{AE+1}{2}\right) + \left(\frac{AE+1}{2}\right)^2$$

$$2(AE) + 1 = (AE)^2 + AE$$

$$(AE)^2 + AE = 2(AE) + 1$$

$$(AE)^2 - AE - 1 = 0$$

Appendix F: Proof of Golden Rectangle

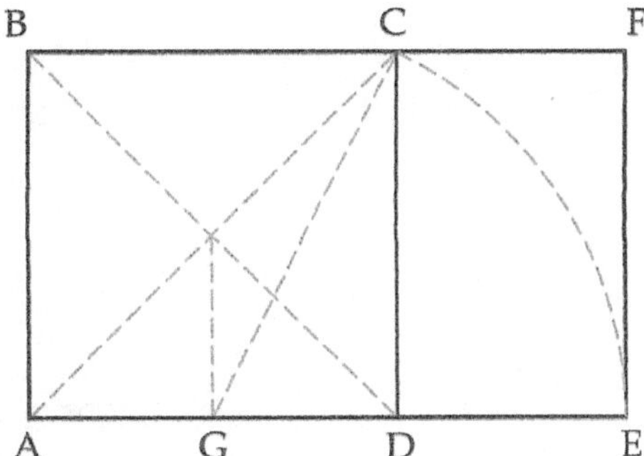

With regard to the rectangle above with side *AB* and base *AE*, point *G* is the center of arc *CGE*, and the midpoint of the base of the square *ABCD*. If *AB* = *AD* = 1, *AG* and *GD* = ½. Using the Pythagorean theorem as related to triangle *GDC*, it can be shown that *AE/AB* = ϕ:

$$(GC)^2 = \left(\frac{1}{2}\right)^2 + 1^2$$

$$GC = \sqrt{\frac{5}{4}}$$

$$= \frac{\sqrt{5}}{2}$$

$$AE = AG + GC$$

$$= \frac{1}{2} + \frac{\sqrt{5}}{2}$$

$$AE = 1.618033989 \ldots = \phi$$

Appendix G: Trigonometric Functions

The trigonometric functions—sine (sin), cosine (cos) and tangent (tan)—relating the sides of a right triangle to the acute angles are expressed as a ratio of the sides *a*, *b* and *c* of the triangle:

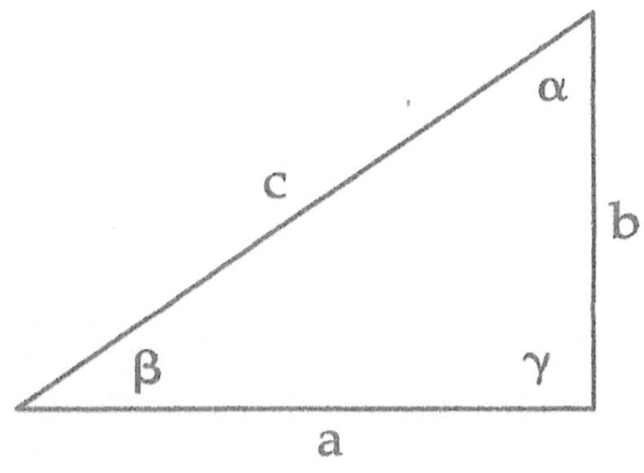

$$\sin \beta = \frac{\textbf{O}pposite\ side}{\textbf{H}ypotenuse} = \frac{b}{c}$$

$$\cos \beta = \frac{\textbf{A}djacent\ side}{\textbf{H}ypotenuse} = \frac{a}{c}$$

$$\tan \beta = \frac{\textbf{O}pposite\ side}{\textbf{A}djacent\ side} = \frac{b}{a}$$

The ratios are easily remembered using the acronym *SOH-CAH-TOA*. The letters of the acronym correspond to the first letter identifying the side of the triangle as related to the first letter of the trigonometric function.

The law of sines relates each angle of any triangle to the side opposite the angle according to proportional ratios:

$$\frac{a}{sin\ \alpha} = \frac{b}{sin\ \beta} = \frac{c}{sin\ \gamma}$$

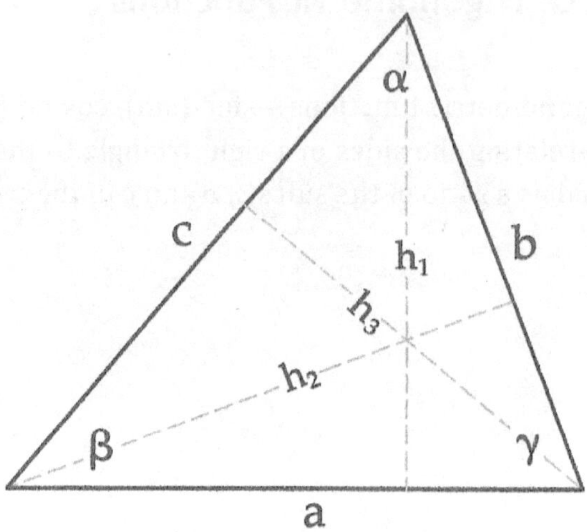

The proportional ratios are derived as follows:

$$\frac{h_1}{b} = \sin \gamma$$

$$h_1 = b \sin \gamma$$

$$\frac{h_1}{c} = \sin \beta$$

$$h_1 = c \sin \beta$$

$$b \sin \gamma = c \sin \beta$$

$$\frac{b}{\sin \beta} = \frac{c}{\sin \gamma}$$

In a similar manner, height h_2 can be extended from apex γ perpendicular to side c to show that $h_2 = a \sin \beta = b \sin \alpha$, such that,

$$\frac{a}{\sin \alpha} = \frac{b}{\sin \beta}$$

Hence:

$$\frac{a}{\sin \alpha} = \frac{b}{\sin \beta} = \frac{c}{\sin \gamma}$$

The derivation of the law of sines can also be determined in another manner involving the area A of the triangle (knowing the area equals one-half the base times the height);

$$A = \frac{1}{2}ah_1 = \frac{1}{2}bh_2 = \frac{1}{2}ch_3$$

Substituting for $h_1 = c\,\sin\beta$, $h_2 = a\,\sin\gamma$ and $h_3 = b\,\sin\alpha$

$$\frac{a \cdot c \cdot \sin\beta}{2} = \frac{b \cdot a \cdot \sin\gamma}{2} = \frac{c \cdot b \cdot \sin\alpha}{2}$$

Multiplying by $\frac{2}{a \cdot b \cdot c}$ results in:

$$\frac{\sin\beta}{b} = \frac{\sin\gamma}{c} = \frac{\sin\alpha}{a}$$

Or the reciprocal,

$$\frac{b}{\sin\beta} = \frac{c}{\sin\gamma} = \frac{a}{\sin\alpha}$$

Appendix H: Polar Coordinates

Polar coordinates locate a point in a plane according to its distance to a central point, and an angle measured from a base line:

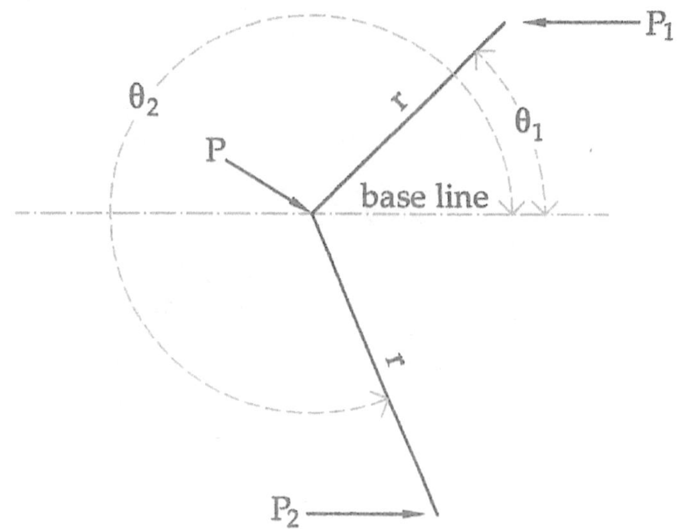

In the drawing above, lines PP_1 and PP_2 are referred to as radius vectors (r_1 and r_2 respectively), having both magnitude and direction. Their directions are determined by angles θ_1 and θ_2 measured in a counterclockwise direction. The angles are measured in terms of degrees (the division of the circumference of a circle into 360 equal arcs) or radians (the *division* of the circumference of a circle in terms of arcs related to the constant pi (π).

Since the circumference of a circle is equal to pi times the diameter, and since the diameter is twice the radius, the circumference is equal to two times pi times the radius ($C = 2\pi r$). If the radius is equal to 1, the circumference is equal to two times pi ($C = 2\pi$). Hence there are 2π radians in a circle.

The following diagram illustrates various angles measured in radians corresponding to the same angles measured in degrees:

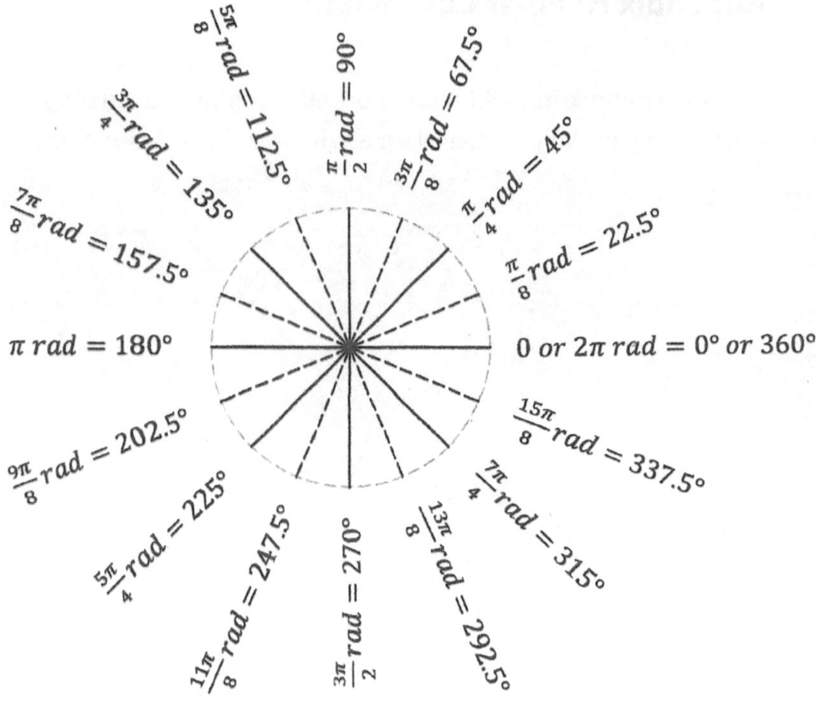

A radian (abbreviated as *rad*) angle is not limited to one sweep or rotation of the circumference of a circle. Any number of rotations is possible. Hence the magnitude of a radius vector relates to the number of times it sweeps past a complete circle. For example, π, *3* π, *5*π, and so forth, all establish the same direction of a radius vector such that its magnitude, as established by a polar equation involving an angle, varies.

Since 2π radians are equal to 360°, one radian is equal to 360° divided by 2π:

$$2\pi\ rad = 360°$$

$$1rad = \frac{360°}{2\pi}$$

$$= 57.2958°$$

Appendix I:
Derivation of the Mathematical Constant "e"

It is beyond the scope of this book to discuss the relevance of the mathematical constant e (2.71828...) except to illustrate that its value is derived by two different methods. The first is the limit of the sum of the reciprocals of all the factorials:

$$e = 1 + \frac{1}{1!} + \frac{1}{2!} + \frac{1}{3!} + \frac{1}{4!} + \ ...$$

$$= 1 + \frac{1}{1} + \frac{1}{1 \cdot 2} + \frac{1}{1 \cdot 2 \cdot 3} + \frac{1}{1 \cdot 2 \cdot 3 \cdot 4} \ ...$$

$$= 2.71828$$

The second is that it is the sum of two plus the limit of the sum of the continuing fraction,

$$e = 2 + \cfrac{1}{1 + \cfrac{1}{2 + \cfrac{1}{1 + \cfrac{1}{1 + \cfrac{1}{4 + \cfrac{1}{1 + \cfrac{1}{1 + \cfrac{1}{6 + \cfrac{1}{1...}}}}}}}}}$$

Which can also be written: 2;1,2,1,1,4,1,1,6,1,1,...

Appendix J: Length of a Logarithmic Spiral

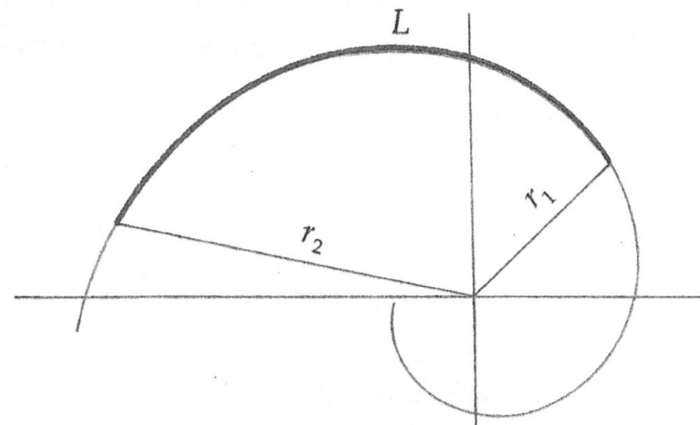

The equation $L = (1/b)\left(\sqrt{1 + b^2}\right)(r_2 - r_1)$, denoting the length of the logarithmic spiral curve from r_1 to r_2 (r_1 and r_2 can have any direction or magnitude) is derived from the general formula for the logarithmic spiral, $r = ae^{b\theta}$. By definition, a, e, and b are constants while r and θ vary. Using calculus, the rate of change of r with respect to θ is determined (note that $a = 1$).

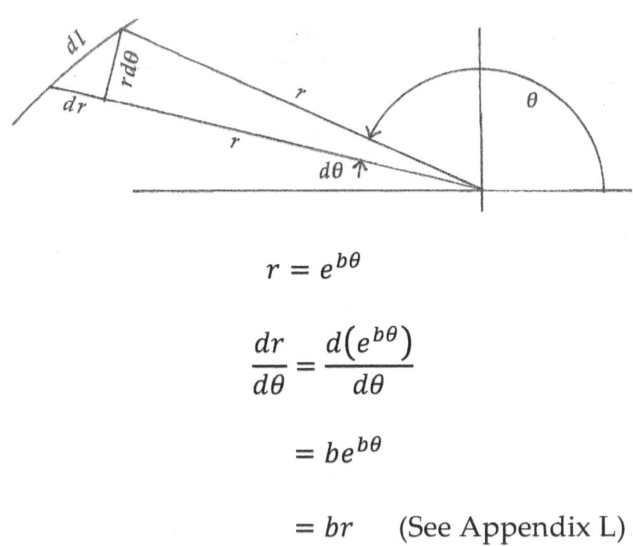

$$r = e^{b\theta}$$

$$\frac{dr}{d\theta} = \frac{d\left(e^{b\theta}\right)}{d\theta}$$

$$= be^{b\theta}$$

$$= br \quad \text{(See Appendix L)}$$

$$dr = brd\theta$$

With regard to the geometry indicated in the previous diagram, from the Pythagorean theorem squaring the sides of a triangle,

$$dl^2 = dr^2 + (rd\theta)^2$$

$$= (brd\theta)^2 + (rd\theta)^2$$

$$dl^2 = b^2 r^2 d\theta^2 + r^2 d\theta^2$$

$$= r^2 d\theta^2 (b^2 + 1)$$

$$= (b^2 + 1)(rd\theta)^2$$

$$dl = \sqrt{b^2 + 1} \cdot (rd\theta)$$

Since $r = ae^{b\theta}$:

$$dl = \sqrt{b^2 + 1} \cdot ae^{b\theta} d\theta$$

$$L = \int_{\theta_1}^{\theta_2} \sqrt{b^2 + 1} \cdot ae^{b\theta} d\theta$$

$$= \frac{1}{b}\sqrt{b^2 + 1} \cdot \left(ae^{b\theta_2} - ae^{b\theta_1}\right)$$

Since $r_i = ae^{b\theta_i}$

$$L = \frac{1}{b}\sqrt{1 + b^2} \cdot (r_2 - r_1)$$

Appendix K: Area of a Logarithmic Spiral

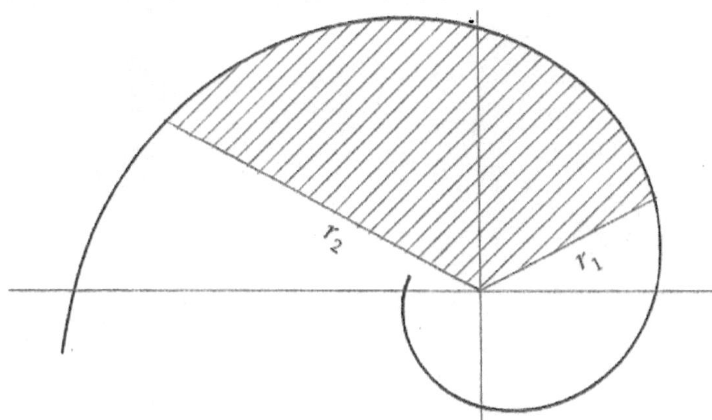

The equation, $A = (1/4b)(r_2^2 - r_1^2)$ denoting the area of the logarithmic spiral curve indicated by the hatching in the diagram above (r_1 and r_2 can have any direction or magnitude) is derived from the general formula for the logarithmic spiral, $r = ae^{b\theta}$.

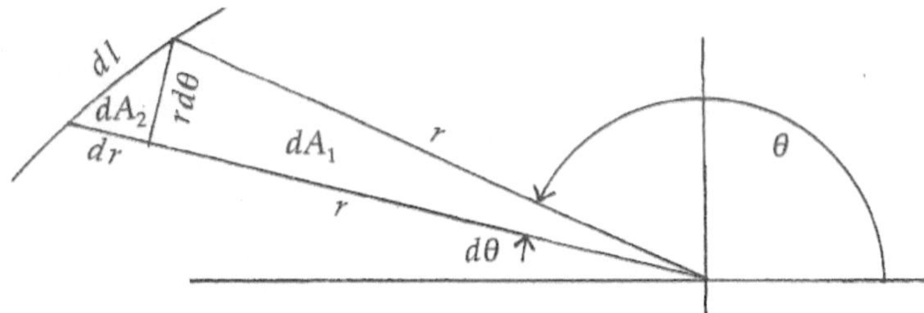

In the diagram above, the area dA_1 of the larger triangle is:

$$dA_1 = \frac{1}{2}r \times rd\theta$$

The area dA_2 of the smaller triangle is:

$$dA_2 = \frac{1}{2}dr \times rd\theta$$

For very small dr and $d\theta$, dA_2 is negligible compared to dA_1, thus:

$$A = \int_{\theta_1}^{\theta_2} dA_1$$

$$= \int_{\theta_1}^{\theta_2} \frac{1}{2} r^2 d\theta$$

For $r(\theta) = ae^{b\theta}$

$$A = \int_{\theta_1}^{\theta_2} \frac{a^2}{2} e^{2b\theta} d\theta$$

$$= \frac{a^2}{4b} \Big|_{\theta_1}^{\theta_2} e^{2b\theta}$$

$$= \frac{1}{4b} \left(a^2 e^{2b\theta_2} - a^2 e^{2b\theta_1} \right)$$

$$= \frac{1}{4b} \left(\left(ae^{b\theta_2}\right)^2 - \left(ae^{b\theta_1}\right)^2 \right)$$

$$A = \frac{1}{4b} \left(r_2^2 - r_1^2 \right)$$

Appendix L:
Slope of a Logarithmic Spiral at any Radius Vector

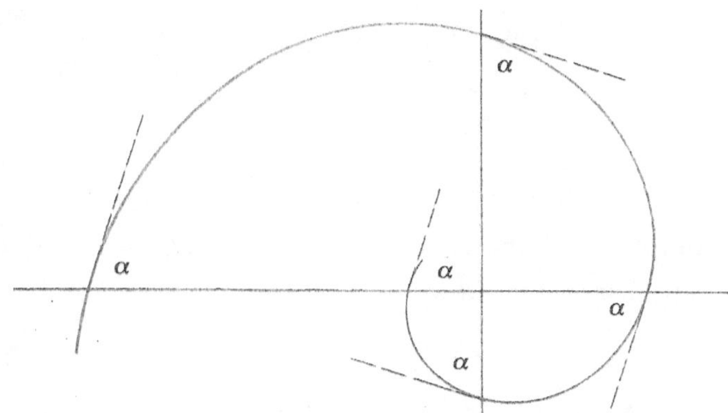

The slope of a logarithmic spiral at any radius r corresponding to tangential angle α (see diagram above) is derived from the general formula for a logarithmic spiral,

$$r = ae^{b\theta}$$

An analysis of the geometry of dr and $d\theta$ in the diagram below,

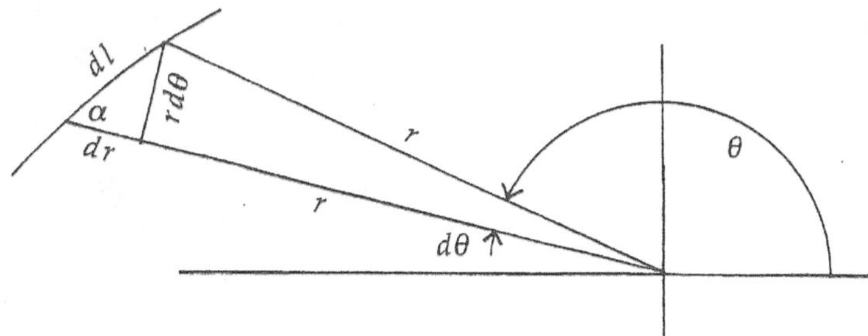

shows that:

$$\frac{dr}{rd\theta} = \cot \alpha$$

Substituting $\frac{dr}{d\theta} = br$ (see appendix J):

$$\frac{br}{r} = \cot \alpha$$

$$b = \cot \alpha$$

$$\frac{1}{b} = \tan \alpha$$

Thus α is determined knowing the value of b. Substituting $b = \cot \alpha$ in the equation, $r = ae^{b\theta}$, and if a = 1:

$$r = e^{(\cot \alpha)\theta}$$

$$= e^{(\cot \alpha)\theta}$$

$$\theta \cot \alpha = \ln$$

$$\cot \alpha = \frac{\ln r}{\theta}$$

$$\tan \alpha = \frac{\theta}{\ln r}$$

Appendix M:
Polar Equation for the Golden Triangle Spiral

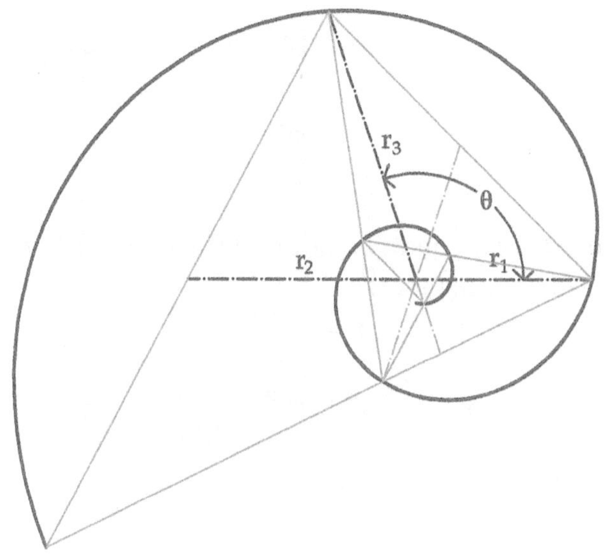

This derivation was made using AutoCAD computer software to determine θ and r_3. If the golden triangles are positioned as shown above, and if r_1 equals 1, $\theta = 108°$ ($3\pi/5$ radians), and $r_3 = \phi$. The dash-dot lines locating the pole of the spiral extend from a base angle to the midpoint of the opposite side. The polar equation for the spiral is derived using the formula, $r = r_2^{\theta/\pi}$ (see page 58).

$$r_3 = r_2^{(3\pi/5)/\pi}$$

$$r_3 = r_2^{3/5}$$

If $r_3 = \phi$, $\qquad\qquad\qquad \phi = r_2^{3/5}$

From the relationship: if $n = a^{1/s}$, then $a = n^s$,

$$r_2 = \phi^{5/3}$$

$$r = r_2^{\theta/\pi}$$

$$= \left(\phi^{5/3}\right)^{\theta/\pi}$$

$$\boldsymbol{r = \phi^{5\theta/3\pi}}$$

The standard polar equation for the spiral in terms of e is determined in a similar manner:

$$r = ae^{b\theta}$$

If $a = 1$ $\qquad\qquad$ $r = e^{b\theta}$

Since $r_3 = \phi$ when $\theta = (3/5)\pi$,

$$\phi = e^{b(3\pi/5)}$$

$$ln\ \phi = b(3\pi/5)$$

$$\left(\frac{5}{3\pi}\right) ln\ \phi = b$$

$$b = 0.254951672$$

$$r = e^{(0.254951672)\theta}$$

Replacing a to establish the size of the spiral:

$$\boldsymbol{r = ae^{(0.25495)\theta}}$$

Appendix N: Polar Equation for Golden Rectangle Spiral

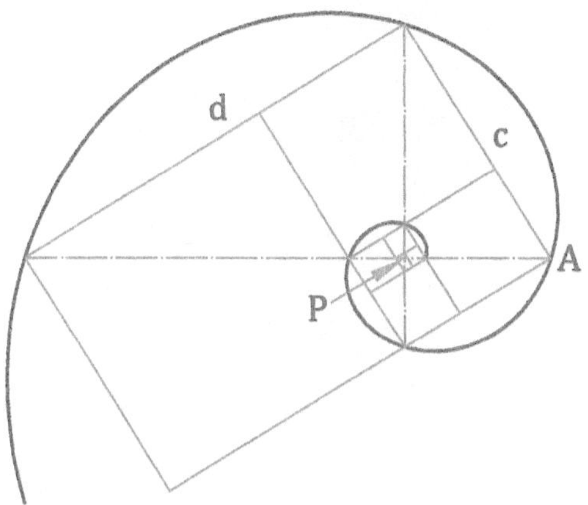

$$\frac{d}{c} = \frac{d'}{c'} = \phi$$

The modified polar equation for the logarithmic spiral generated from rotating golden rectangles is derived as follows:

If $PA = 1$[*]
$$r = \left(\frac{d'}{c'}\right)^{2\theta/\pi}$$

$$r = \phi^{2\theta/\pi}$$

Also if $c = 1$ and $d = \phi$,

$$r = \frac{\phi^{2\theta/\pi}}{\sqrt{1+\phi^2}}$$

The standard polar equation for the spiral in terms of e is determined in the following manner:

$$\phi^{2\theta/\pi} = ae^{b\theta}$$

If $a = 1$,
$$\ln\left(\phi^{2\theta/\pi}\right) = b\theta$$

---

[*] See "Logarithmic Spirals as Related to Rectangles"

If $\theta = \pi$
$$ln\left(\phi^{2\pi/\pi}\right) = b\pi$$

$$ln(\phi^2)\frac{1}{\pi} = b$$

$$\frac{ln\phi^2}{\pi} = b = 0.30635\ldots$$

$$r = ae^{(0.30635\ldots)\theta}$$

Appendix O: Logarithmic Spiral Tangent at the Corner of a Rectangle

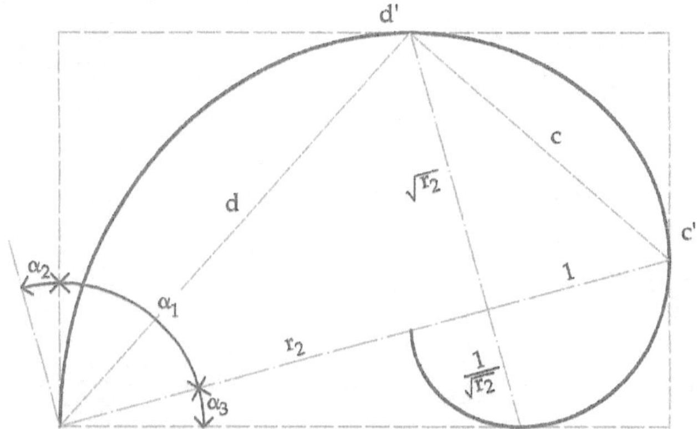

In the drawing above, tangential angle α_1, (where the spiral intersects the corner of the rectangle) is derived in the following manner. From Appendix L, $\cot \alpha_1 = \frac{\ln(r/a)}{\theta}$. If $a = 1$ and $\theta = \pi$:

$$\cot \alpha_1 = \frac{\ln r_2}{\pi}$$

$$\tan \alpha_1 = \frac{\pi}{\ln r_2}$$

$$\tan \alpha_3 = \frac{1/\sqrt{r_2}}{r_2}$$

$$\tan \alpha_3 = \frac{1}{r_2 \sqrt{r_2}}$$

Hence,
$$\alpha_3 = \tan^{-1}(1/r_2\sqrt{r_2})$$

$$\alpha_2 = 90° - \alpha_1$$

In terms of radians:
$$\alpha_2 = \frac{\pi}{2} - \alpha_1$$

Since $\alpha_2 = \alpha_3$:
$$\alpha_3 = \frac{\pi}{2} - \alpha_1$$

Substituting $\alpha_3 = tan^{-1}\left(\frac{1}{r_2\sqrt{r_2}}\right)$, and $\alpha_1 = tan^{-1}\left(\frac{\pi}{lnr_2}\right)$:

$$tan^{-1}\left(\frac{1}{r_2\sqrt{r_2}}\right) = \frac{\pi}{2} - tan^{-1}\left(\frac{\pi}{lnr_2}\right)$$

Substituting various values of r_2 on each side of the above equation until it balances, reveals that if $r_2 = 2.368064 \ldots$,

$$tan^{-1}\left(\frac{1}{3.6441}\right) = \frac{\pi}{2} - tan^{-1}\left(\frac{\pi}{0.8621}\right)$$

$$tan^{-1}(0.2744) = \frac{\pi}{2} - tan^{-1}(0.8621)$$

$$0.2678 \, rad = 1.5708 \, rad - 1.3030 \, rad$$

$$0.2678 \, rad = 0.2678 \, rad$$

Hence:
$$r = 2.3681^{\theta/\pi}$$

Or in terms of e where $b = (ln\ 2.3681)/\pi$,

$$r = ae^{0.2744\theta}$$

Substituting $r_2 = 2.3681$, and $\theta = \pi$ in the equation $\tan \alpha_1 = \theta/ln\ r_2$:

$$\tan \alpha_1 = \frac{\pi}{\ln 2.3681}$$

$$\alpha_1 = \tan^{-1} 3.6442$$

$$\boldsymbol{\alpha_1 = 74.6551°}$$

$$\alpha_3 = 90° - 74.6551$$

$$\boldsymbol{\alpha_3 = 15.3443°}$$

$$\frac{d'}{2.3681 + 1} = \cos \alpha_3$$

$$\boldsymbol{d' = 3.2482}$$

$$d = \sqrt{2.3681^2 + \left(\sqrt{2.3681}\right)^2}$$

$$\boldsymbol{d = 2.8242}$$

$$c = \sqrt{1^2 + \left(\sqrt{2.3681}\right)^2}$$

$$\boldsymbol{c = 1.8352}$$

$$\frac{c}{d} = \frac{c'}{d'}$$

$$c' = \frac{cd'}{d}$$

$$\boldsymbol{c' = 2.1107}$$

$$\boldsymbol{\frac{d'}{c'} = \frac{d}{c} = 1.5389}$$

Appendix P: Construction of a Pentagon

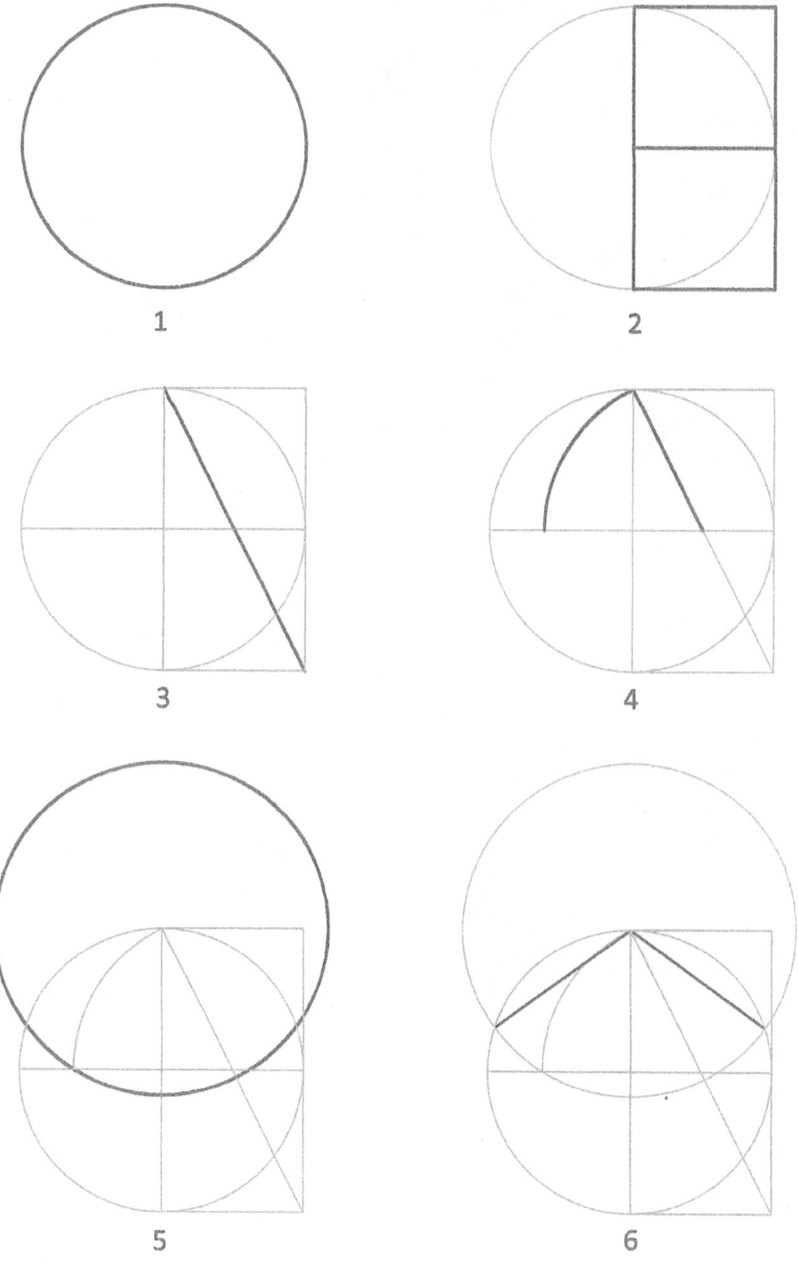

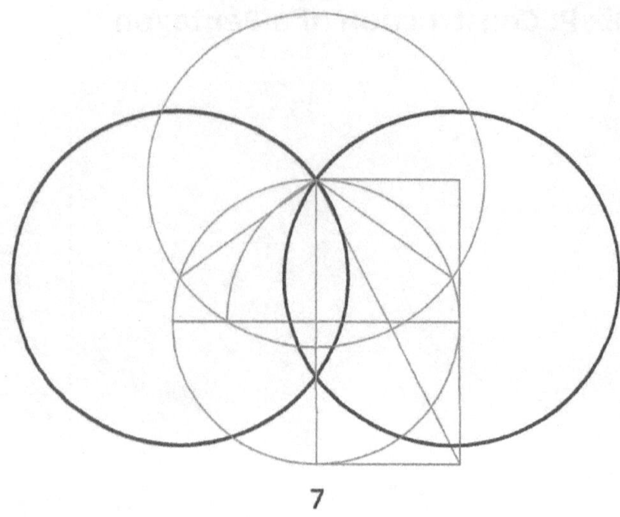

7

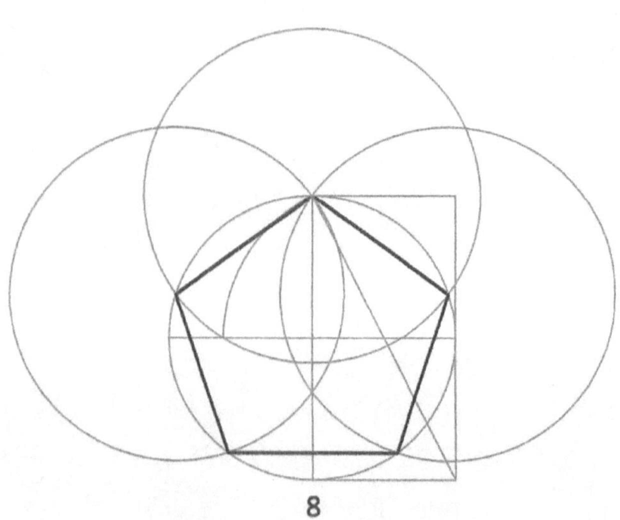

8

Appendix Q: Construction of Egg Profiles

After discovering that the proportions of the eggs of birds (diameter to length) appear to be related to the sequence, 1/2, 2/3, 3/4, 4/5, 5/6, …, the thought came to me that perhaps the silhouette of an egg can be replicated geometrically. My approach was to use arcs of circles: a logical approach given that the girth of every egg I observed is a perfect circle. The following drawings (unique to this publication) illustrate the geometry based on the proportions of the rectangle, and the diameters of three different circles:

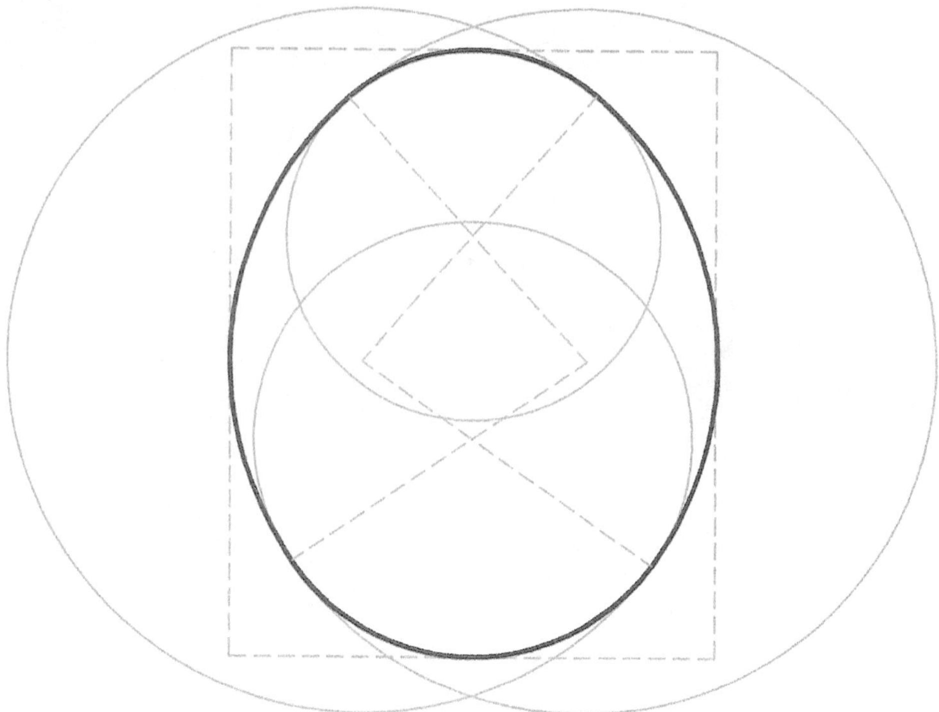

Ostrich egg, 4:5 rectangle

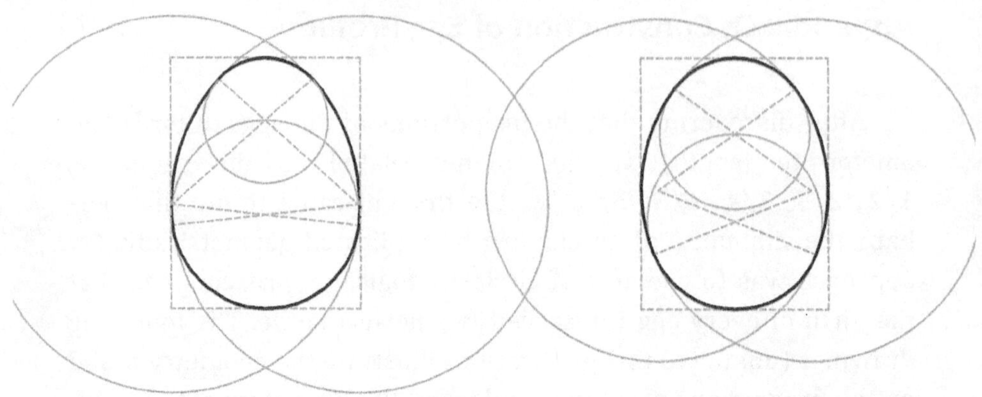

Chicken eggs, 3:4 rectangles. The shape of the egg profile is related to the size and placement of the circles.

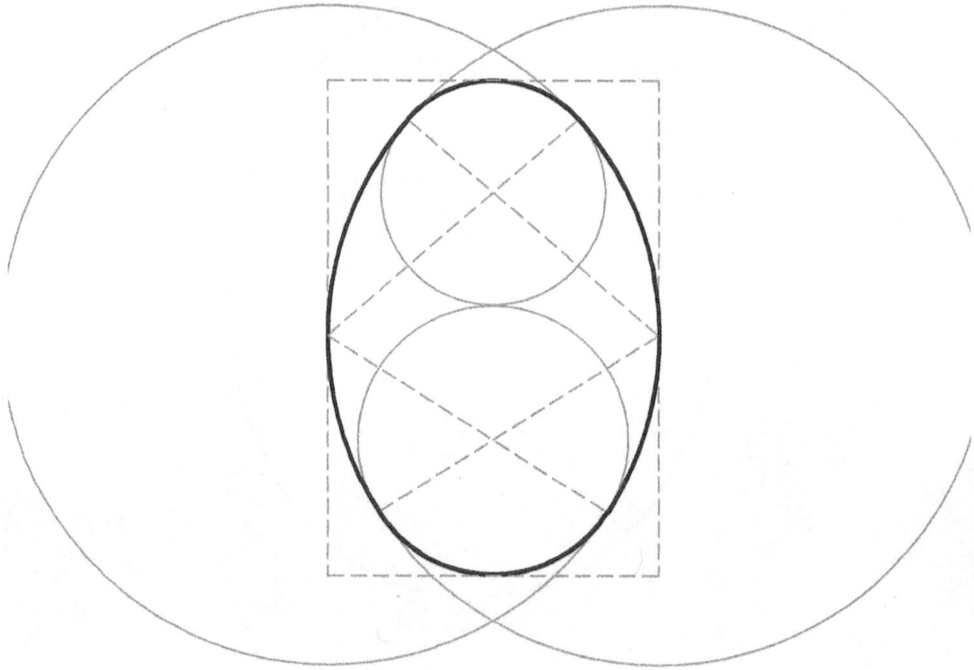

Emu egg, 2:3 rectangle

Appendix R: Alternate spiral constructions

The magnitudes of the radius vectors required for the construction of a spiral based on sixteen equally spaced radius vectors can be determined knowing the proportionality between them. In the diagram below, a value is assigned to radius vectors r_2. Radius vector r_1 can be any value less than r_2. However, for simplicity and conformation to mathematical convention, r_1 is given the value of one. The subscript number at each remaining radius vector indicates the order in which it is determined and plotted.

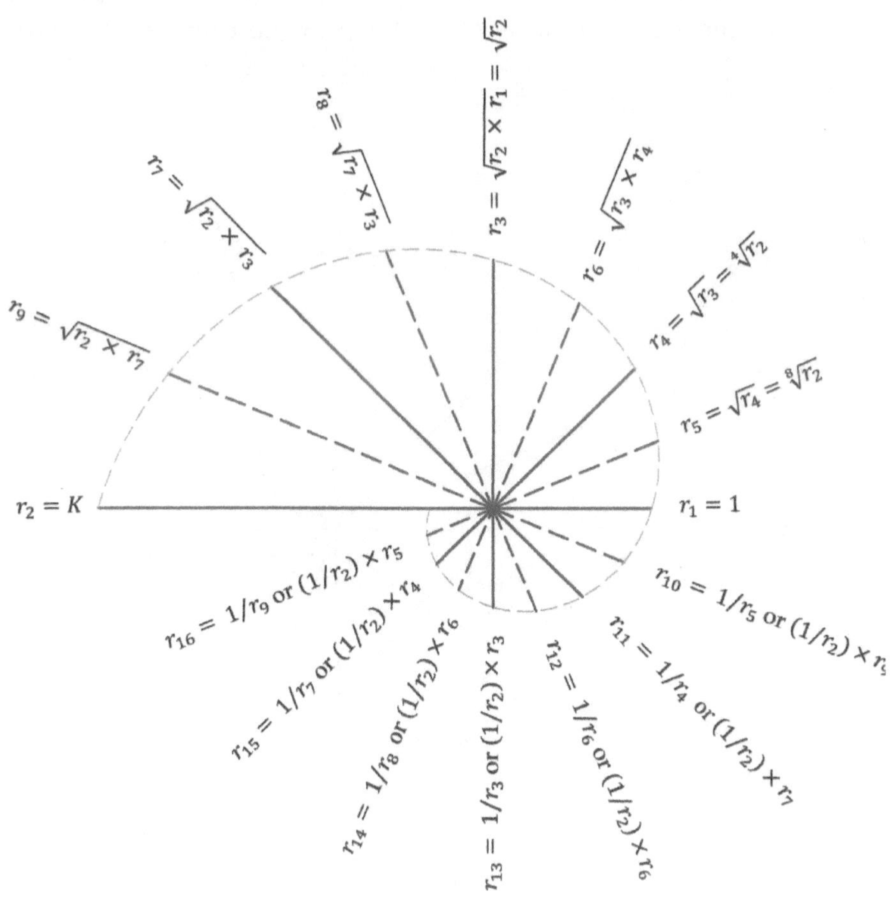

Each radius vector can of course be determined using the standard equation for a spiral, $r = ae^{b\theta}$. As has been demonstrated in previous examples, if a is given the value of one, the value of b is determined by specifying a value for r_2, say K, at $\theta = \pi$ radians:

$$K = e^{b\pi}$$

$$\ln K = b\pi$$

$$b = (\ln K)/\pi$$

$$r = e^{(\ln K)\theta/\pi}$$

The radius vectors are plotted in the diagram below. Except for r_1 and r_2, the order is different from the previous example.

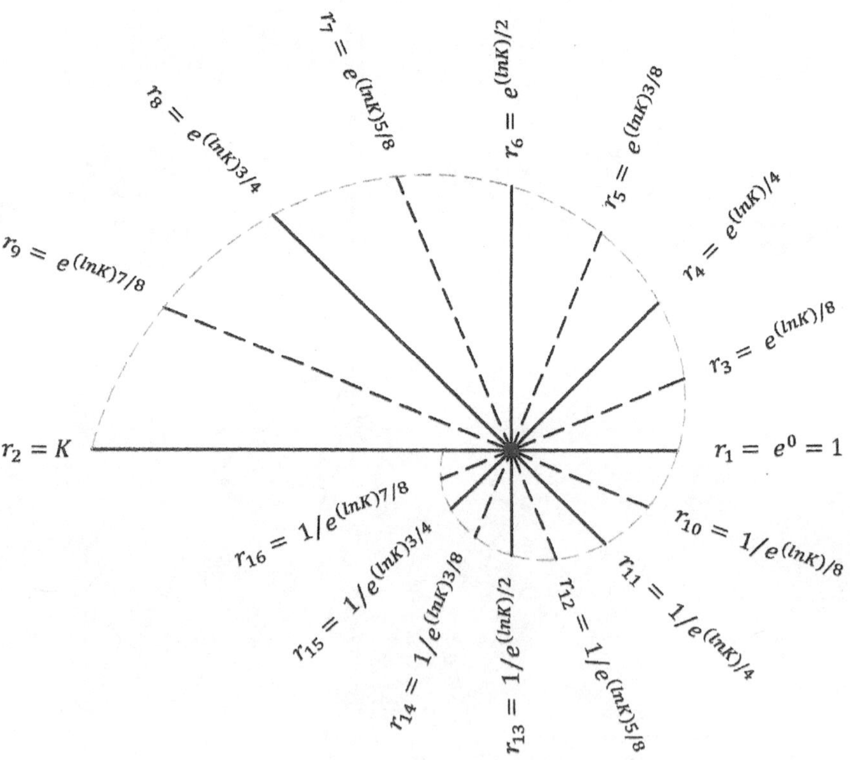

Appendix S:
Derivation of the Length of the Koch Curve

The length of the side of the starting triangle of the Koch curve is a. Hence the perimeter L_0 is $3 \times a$.

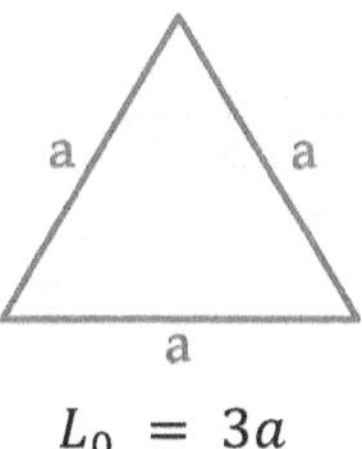

$$L_0 = 3a$$

The length of the sides of the added triangles of the first iteration is one-third of a, hence the new perimeter is:

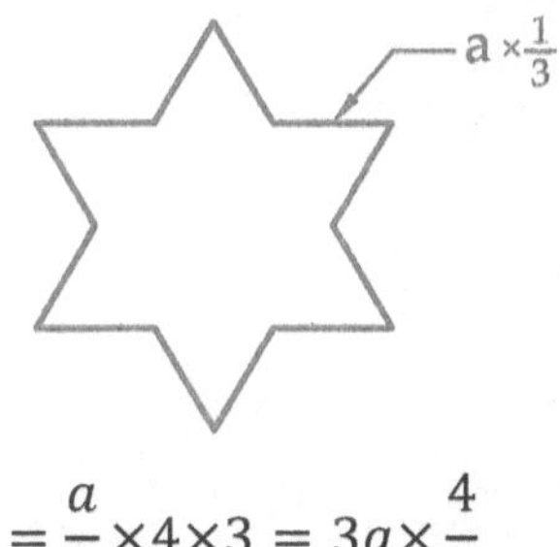

$$L_1 = \frac{a}{3} \times 4 \times 3 = 3a \times \frac{4}{3}$$

The length of the side of the added triangles, $a \times 1/3 \times 1/3$, of the second iteration is one-third of the length of the side of the first iteration. Hence the new perimeter of the second iteration is:

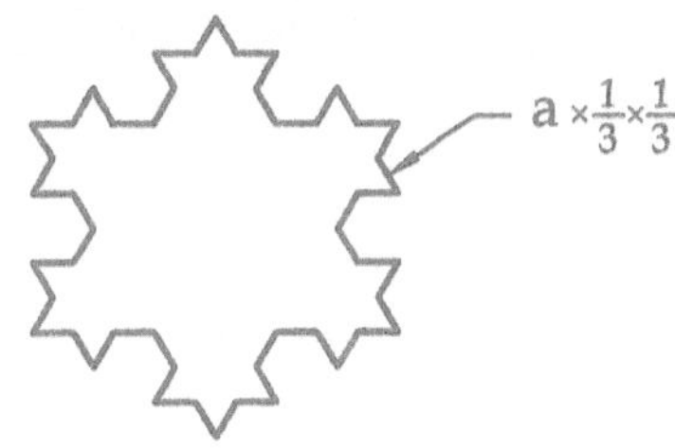

$$L_2 = \frac{a}{3} \times \frac{1}{3} \times 16 \times 3 = \frac{a \times 3 \times 16}{3^2} = 3a \times \frac{4^2}{3^2}$$

The length of the sides of the added triangles of the third iteration, $a \times 1/3 \times 1/3 \times 1/3$, is one-third of the length of the side of the second iteration, $a \times 1/3 \times 1/3$, hence the new perimeter is:

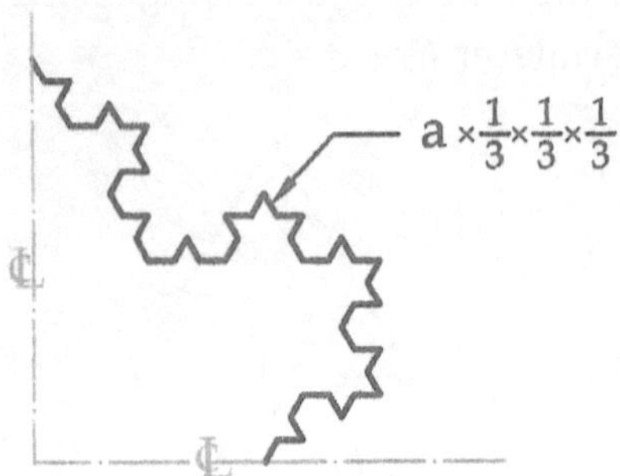

The diagram illustrates one quadrant of the curve

$$L_3 = \frac{a}{3^2} \times \frac{1}{3} \times 48 \times 4 = \frac{a \times 3 \times 16 \times 4}{3^3} = 3a \times \frac{4^3}{3^3}$$

The perimeters of each of the iterations form a geometric progression:

$$3a, \quad 3a \times \frac{4}{3}, \quad 3a \times \left(\frac{4}{3}\right)^2, \quad 3a \times \left(\frac{4}{3}\right)^3, \quad 3a \times \left(\frac{4}{3}\right)^4 \dots$$

Hence, the length of the curve at any iteration n is:

$$L_n = L_0 \times \left(\frac{4}{3}\right)^n$$

Appendix T: Area of the Koch Curve

The area of the initial equilateral triangle is A_0:

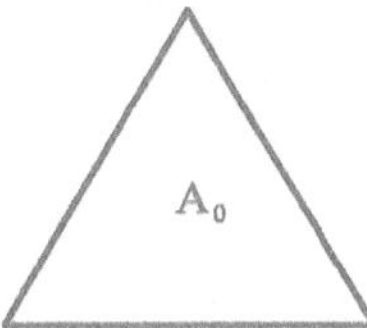

The first iteration adds three triangles, each one-ninth the area of the original triangle:

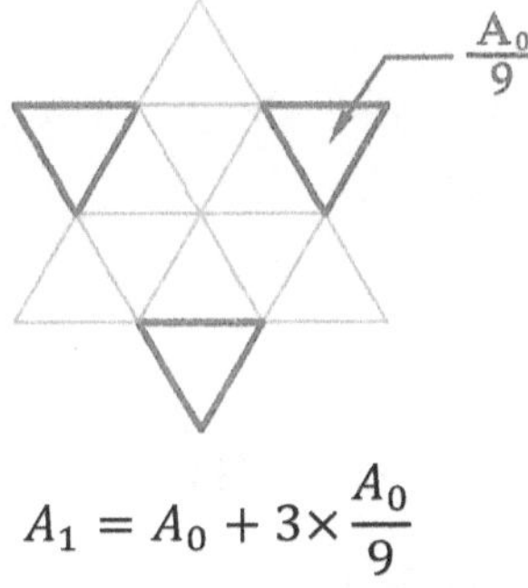

$$A_1 = A_0 + 3 \times \frac{A_0}{9}$$

The second iteration adds two more triangles to each of the previously added triangles. The area of each of these (for a total of twelve) is one-ninth the area of the previously added triangle.

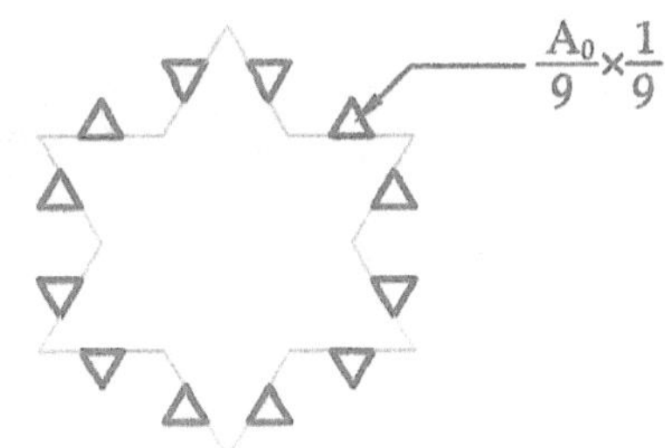

$$A_2 = A_0 + 3 \times \frac{A_0}{9} + 12 \times \frac{A_0}{9^2}$$

The third iteration adds a total of 12 triangles to the perimeter of each quadrant as indicated in the drawing below. The area of each added triangle is one-ninth the area of each triangle added in the second iteration. The total number of triangles added in the third iteration is $12 \times 4 = 48$:

$$\frac{A_0}{9} \times \frac{1}{9} \times \frac{1}{9}$$

$$A_3 = A_0 + 3 \times \frac{A_0}{9} + 12 \times \frac{A_0}{9^2} + 48 \times \frac{A_0}{9^3}$$

$$A_{n \to \infty} = A_0 + \frac{3A_0}{9} + \frac{3A_0 \times 4}{9^2} + \frac{3A_0 \times 16}{9^3} + \cdots$$

$$= A_0 \left[1 + 3 \left(\frac{1}{9} + \frac{4}{9^2} + \frac{4^2}{9^3} + \cdots \right) \right]$$

$$= A_0 \left[1 + 3 \times \frac{4}{4} \times \left(\frac{1}{9} + \frac{4}{9^2} + \frac{4^2}{9^3} + \cdots \right) \right]$$

$$= A_0 \left[1 + \frac{3}{4} \times 4 \left(\frac{1}{9} + \frac{4}{9^2} + \frac{4^2}{9^3} + \cdots \right) \right]$$

$$= A_0 \left[1 + \frac{3}{4} \times \left(\left(\frac{4}{9} \right)^1 + \left(\frac{4}{9} \right)^2 + \left(\frac{4}{9} \right)^3 + \cdots \right) \right]$$

$$\left(\left(\frac{4}{9} \right)^1 + \left(\frac{4}{9} \right)^2 + \left(\frac{4}{9} \right)^3 + \cdots \right) = \frac{4}{5} \qquad \text{(See appendix U.)}$$

$$A_{n \to \infty} = A_0 \left[1 + \frac{3}{4} \times \frac{4}{5} \right]$$

$$A_{n \to \infty} = A_0 \times \frac{8}{5}$$

Appendix U: Sum of a Geometric Series

The sum of the infinite series $a + a^2 + a^3 + \cdots + a^n + \cdots a^\infty$ where a is less than 1 and greater than 0, is derived as follows:

$$a + a^2 + a^3 + \cdots + a^n + \cdots a^\infty = a(1 + a + a^2 + a^3 + \cdots + a^n + a^\infty)$$

$$\sum_{n=1}^{\infty} a^n = a\left(1 + \sum_{n=1}^{\infty} a^n\right)$$

$$= a + a \sum_{n=1}^{\infty} a^n$$

$$\sum_{n=1}^{\infty} a^n - a \sum_{n=1}^{\infty} a^n = a$$

$$\sum_{n=1}^{\infty} a^n (1 - a) = a$$

$$\sum_{n=1}^{\infty} a^n = \frac{a}{1 - a}$$

Hence where $a = 4/9$ (previous page):

$$\sum_{n=1}^{\infty} a^n = \frac{4/9}{1 - 4/9} = \frac{4/9}{5/9} = \frac{4}{5}$$

Appendix V: Right Triangles as Related to the Diagonal of a Rectangle and its Associated Logarithmic Spiral

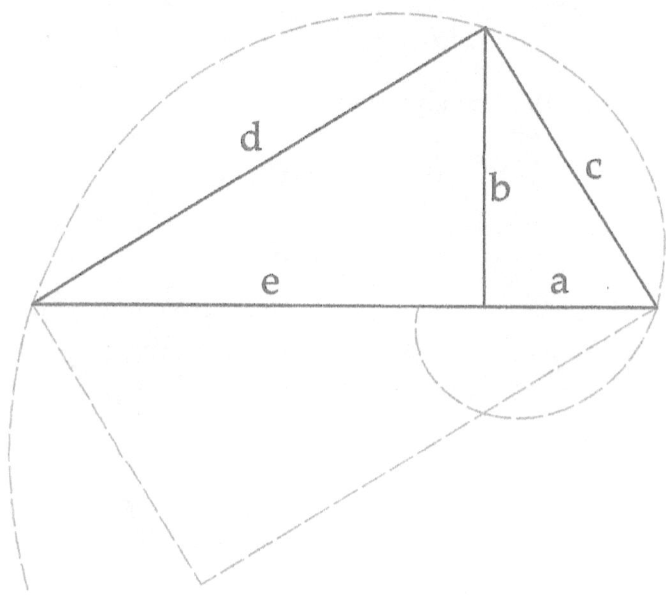

The derivation of a in terms of c and d in the diagram above is determined as follows:

$$\frac{d}{e + a} = \frac{e}{d}$$

$$\frac{d^2}{e + a} = e$$

$$d^2 + c^2 = (e + a)^2$$

$$\sqrt{d^2 + c^2} = e + a$$

Substituting for $e + a$ in the first equation above:

$$\frac{d}{\sqrt{d^2 + 2}} = \frac{e}{d}$$

$$\frac{d^2}{\sqrt{d^2 + c^2}} = e$$

Continuing with the first equation:

$$(e + a)e = d^2$$

$$e + a = \frac{d^2}{e}$$

$$a = \frac{d^2}{e} - e$$

$$a = \frac{d^2}{\dfrac{d^2}{\sqrt{d^2 + c^2}}} - \frac{d^2}{\sqrt{d^2 + c^2}}$$

$$a = \sqrt{d^2 + c^2} - \frac{d^2}{\sqrt{d^2 + c^2}}$$

$$a\sqrt{d^2 + d^2} = \sqrt{d^2 + c^2}\left(\sqrt{d^2 + c^2} - \frac{d^2}{\sqrt{d^2 + c^2}}\right)$$

$$a\sqrt{d^2 + c^2} = d^2 + c^2 - d^2$$

$$a = \frac{c^2}{\sqrt{d^2 + c^2}}$$

Appendix W: Perimeter of the Pentagram Koch curve

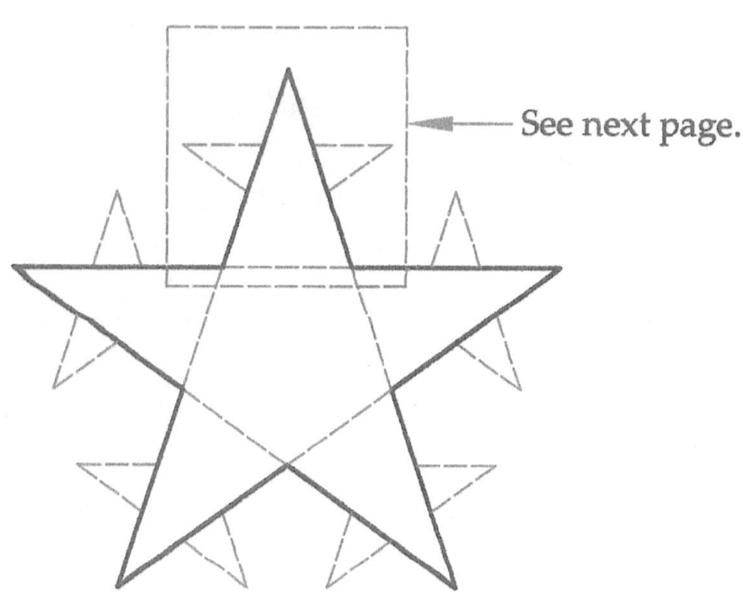

First Iteration

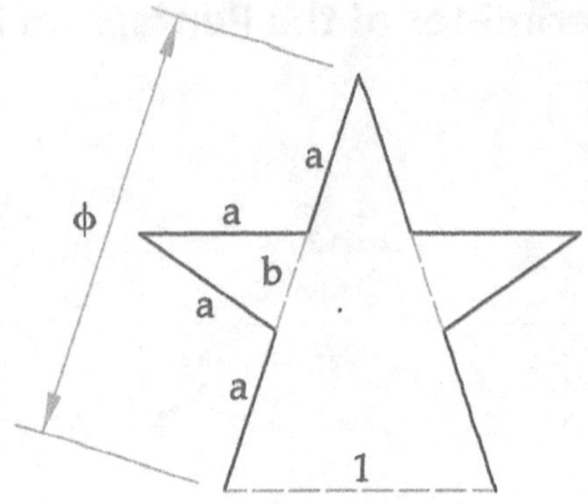

Regarding the first iteration shown in the drawing above,

$$\frac{a}{\phi} = \frac{b}{1}$$

$$2a + b = \phi$$

Hence, $\qquad 2a = \phi - b$

Substituting $b = a/\phi$, $\qquad 2a = \phi - a/\phi$

$$2a + \frac{a}{\phi} = \phi$$

$$a\left(2 + \frac{1}{\phi}\right) = \phi$$

$$a = \frac{\phi}{2 + 1/\phi}$$

Since $\frac{1}{\phi} = \phi - 1$ $\qquad a = \frac{\phi}{2+\phi-1}$

$$a = \frac{\phi}{1 + \phi}$$

Since $\phi^2 = \phi + 1$ $\qquad a = \frac{\phi}{\phi^2}$

$$a = \frac{1}{\phi}$$

The ratio "R" of the extended length to the original length is:

$$R = \frac{4a}{\phi}$$

$$= \frac{4\left(\frac{1}{\phi}\right)}{\phi}$$

$$R = \frac{4}{\phi^2}$$

There are always ten sides to the star—two for each point. Therefore, the total length (as related to the iterations) is always ten times a side length. Each time the iteration is done, each side increases in length, with the new length being the previous length times R. If the number of iterations is n, and the initial side length is phi, then the perimeter length is:

$$P = 10\phi R^n$$

$$= 10\phi \left(\frac{4}{\phi^2}\right)^n$$

Hence the length (perimeter) of the Koch curve of the pentagram expands according to the geometric progression,

$$10\phi, \ 10\phi\left(\frac{4}{\phi^2}\right), \ 10\phi\left(\frac{4}{\phi^2}\right)^2, \ 10\phi\left(\frac{4}{\phi^2}\right)^3, \ 10\phi\left(\frac{4}{\phi^2}\right)^4, \ ...$$

Appendix X: Area of the Pentagram Koch Curve

The area of the Koch pentagram curve is equal to the area of each iteration plus the areas of the initial pentagon and the five golden triangles. The area of the initial pentagon is assigned the value A_0, which is composed of golden triangle A and equal triangles B_1 and B_2:

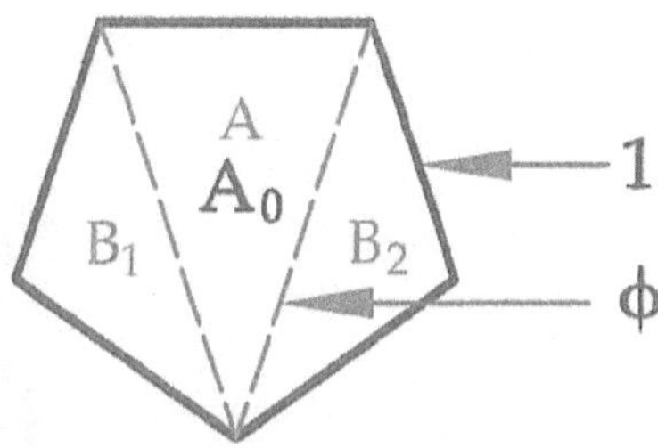

The area of each golden triangle is assigned the value A:

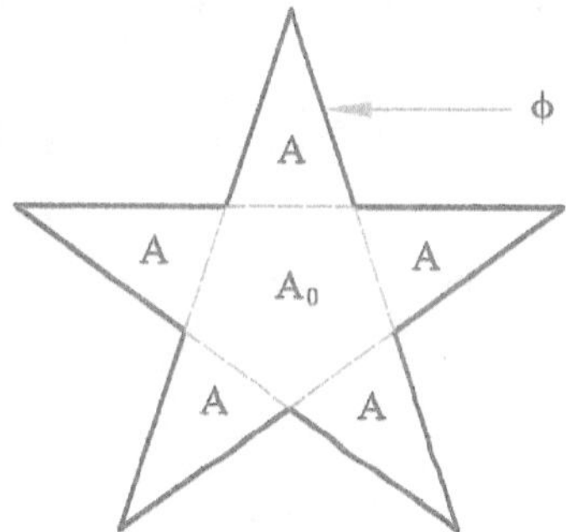

The area (A_1) of each triangle of the first iteration is $A \times \dfrac{1}{\phi^4}$:

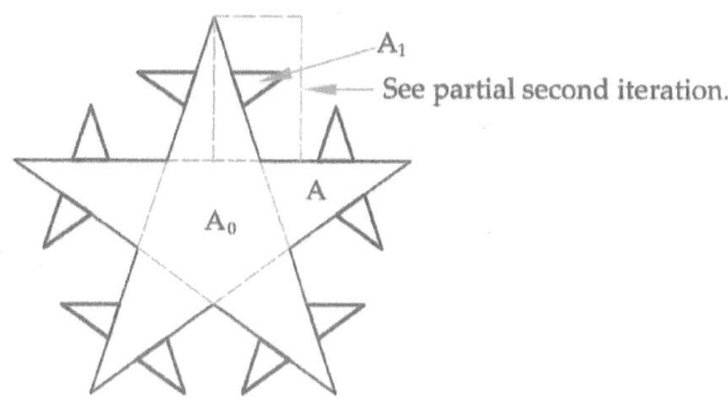

As can be seen, there are ten added triangles. Hence the total added area of the first iteration is $10 \times A \times \frac{1}{\phi^4}$. The area of each triangle of the second iteration is $A_1 \times \frac{1}{\phi^4} = \frac{A}{\phi^4 \times \phi^4} = \frac{A}{\phi^8}$

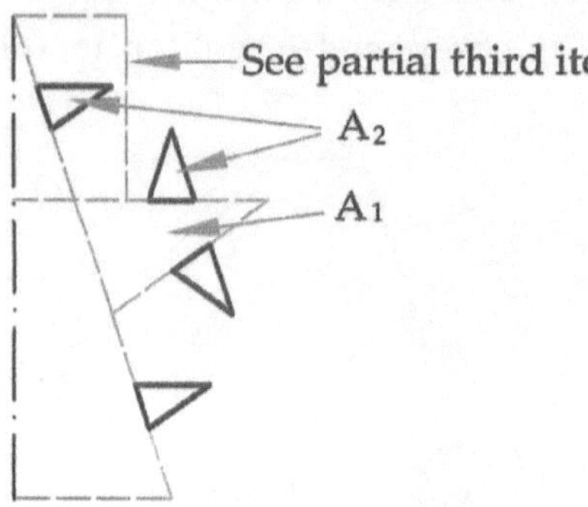

Partial Second Iteration

The total added area of the second iteration includes four triangles for each of the ten triangles of area A_1:

$$10 \times 4 \left(\frac{1}{\phi^8} \right)$$

The area of each triangle of the third iteration is $A_2 \times \frac{1}{\phi^4} = \frac{A}{\phi^8 \times \phi^4} = \frac{A}{\phi^{12}}$:

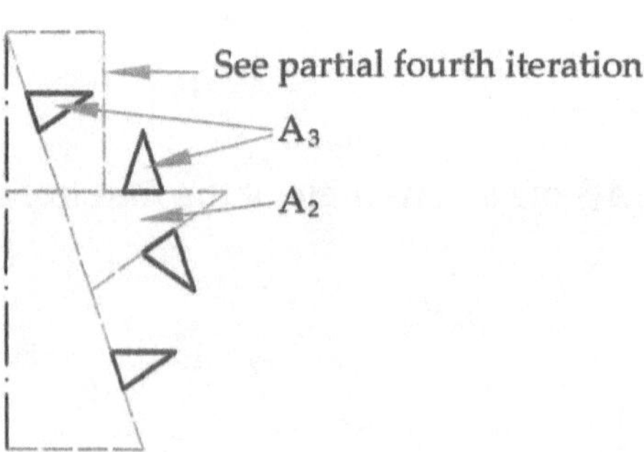

Partial Third Iteration

The total added area of the third iteration includes four triangles for each of the forty triangles of area A_2:

$$40 \times 4 \left(\frac{A}{\phi^{12}} \right)$$

The area of each triangle of the forth iteration is $A_3 \times \frac{1}{\phi^4} = \frac{A}{\phi^{12} \times \phi^4} = \frac{A}{\phi^{16}}$:

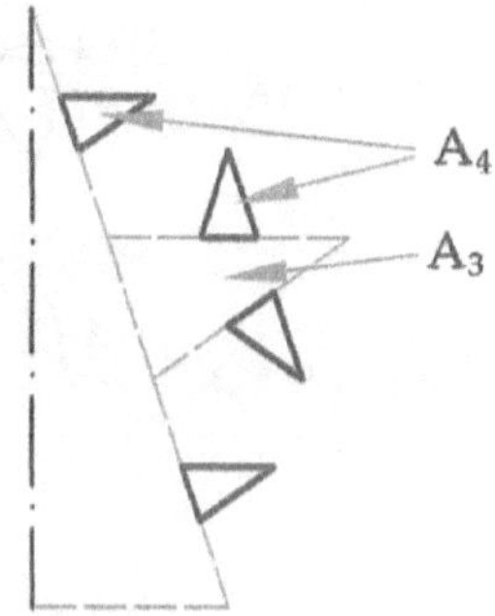

Partial Fourth Iteration

The total added area of the fourth iteration is 4 triangles for each of the 160 triangles of area A_2:

$$160 \times 4 \left(\frac{A}{\phi^{12}} \right)$$

The iterations can continue in a like manner such that the added area at each iteration can be determined. The formula for the area after four iterations, where A_0 is the area of the initial pentagon, and $5A$ is the area of the five golden triangles forming the pentagram, is:

$$A_4 = A_0 + 5A + 10A \left(\frac{1}{\phi^4} + \frac{4}{\phi^8} + \frac{4^2}{\phi^{12}} + \frac{4^3}{\phi^{16}} \right)$$

The area after an infinite number of iterations is determined as follows:

$$A_{n\to\infty} = A_0 + 5A + 10A \times \frac{4}{4}\left(\frac{1}{\phi^4} + \frac{4}{\phi^8} + \frac{4^2}{\phi^{12}} + \frac{4^3}{\phi^{16}} \cdots\right)$$

$$A_{n\to\infty} = A_0 + 5A + 10A \times \frac{1}{4}\left(\frac{4}{\phi^4} + \frac{4^2}{\phi^8} + \frac{4^3}{\phi^{12}} + \frac{4^4}{\phi^{16}} \cdots\right)$$

$$A_{n\to\infty} = A_0 + 5A + 10A \times \frac{1}{4}\left(\left(\frac{4}{\phi^4}\right)^1 + \left(\frac{4}{\phi^4}\right)^2 + \left(\frac{4}{\phi^4}\right)^3 + \left(\frac{4}{\phi^4}\right)^4 \cdots\right)$$

$$A_{n\to\infty} = A_0 + 5A + \frac{10A}{4}\left(\sum_{n=1}^{\infty} \frac{4}{\phi^4}\right)$$

$$= A_0 + 5A + \frac{10A}{4}\left(\frac{\frac{4}{\phi^4}}{1 - \frac{4}{\phi^4}}\right)$$

$$= A_0 + 5A + \frac{10A}{4}\left(\frac{4}{\phi^4 - 4}\right)$$

$$= A_0 + 5A + \frac{10A}{\phi^4 - 4}$$

With regard to the initial pentagon whose area A_0 is equal to area A plus two times area B, it can be shown (see following pages) that $A = \frac{\sqrt{\phi+3/4}}{2}$ and $B = \frac{\phi\sqrt{3-\phi}}{4}$. Hence,

$$A_0 = \frac{\sqrt{\phi + 3/4}}{2} + \frac{\phi\sqrt{3 - \phi}}{2}$$

And therefore,

$$A_{n\to\infty} = \frac{\sqrt{\phi + 3/4}}{2} + \frac{\phi\sqrt{3 - \phi}}{2} + \frac{5\sqrt{\phi + 3/4}}{2} + \frac{10\sqrt{\phi + 3/4}}{2(\phi^4 - 4)}$$

$$A_{n \to \infty} = \frac{1}{2}\left(6\sqrt{\phi + 3/4} + \phi\sqrt{3 - \phi} + \frac{10\sqrt{\phi + 3/4}}{\phi^4 - 4} \right)$$

Thus, as a result of infinite iterations, although the perimeter is infinite, the area is finite.

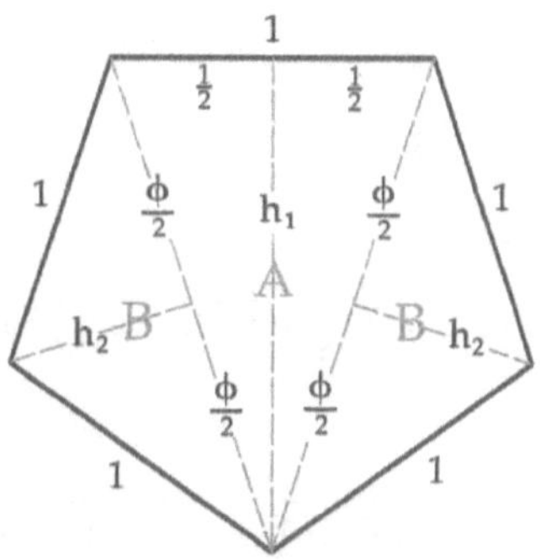

Original pentagon with area designated A_0

With reference to the diagram above, the derivation of area A, for substitution in $A_0 = A + 2B$ (see previous page), is as follows:

$$(h_1)^2 + \left(\frac{1}{2}\right)^2 = \phi^2$$

$$h_1 = \sqrt{\phi^2 - 1/4}$$

$$h_1 = \sqrt{\phi + 1 - 1/4}$$

$$h_1 = \sqrt{\phi + 3/4}$$

$$A = \frac{1}{2} \times 1 \times h_1$$

$$A = \frac{1}{2}\sqrt{\phi + 3/4}$$

The derivation of area B, for substitution in $A_0 = A + 2B$, is as follows:

$$(h_2)^2 + \left(\frac{\phi}{2}\right)^2 = 1^2$$

$$h_2 = \sqrt{1 - \phi^2/4}$$

$$B = \frac{1}{2} \times \phi \times h_2$$

$$B = \frac{1}{2} \times \phi \times \sqrt{1 - \phi^2/4}$$

$$= \frac{\phi}{2} \sqrt{1 - (\phi + 1)/4}$$

$$= \frac{\phi}{2} \sqrt{1 - \phi/4 + 1/4}$$

$$= \frac{\phi}{2} \sqrt{3/4 - \phi/4}$$

$$= \frac{\phi}{2} \sqrt{\frac{1}{4}(3 - \phi)}$$

$$= \frac{\phi}{2} \times \frac{1}{2} \sqrt{3 - \phi}$$

$$\boldsymbol{B = \frac{\phi}{4} \sqrt{3 - \phi}}$$